The Braking Point

America's Energy Dreams *and* Global Economic Realities

The Braking Point

America's Energy Dreams *and* Global Economic Realities

MARK A. STANSBERRY
with **JASON P. REIMBOLD**

HAWK
PUBLISHING
GROUP

LIBRARY OF CONGRESS CATALOG IN PUBLICATION DATA

The Braking Point: America's Energy Dreams and Global Economic Realities / Mark A. Stansberry with Jason P. Reimbold

Hawk Publishing Group Edition 2008
Copyright © 2008 by Mark A. Stansberry with Jason P. Reimbold

All rights reserved under International and Pan-American Copyright Conventions. No part of this book may be reproduced, stored in a retrieval system or transmitted in any form by an electronic, mechanical, photocopying, recording means or otherwise, without prior written permission of the author.

Cover design: nbishopsdesigns@cox.net
Jason Reimbold's author photo:
Andy's Fine Portraits, Tulsa, Oklahoma

ISBN 1-930709-67-6
Library of Congress Control Number: 2007940455

Published in the United States by HAWK Publishing Group

HAWK Publishing Group
7107 South Yale Avenue #345
Tulsa, OK 74136
918-492-3677
www.hawkpub.com

HAWK and colophon are trademarks belonging to the HAWK Publishing Group. Printed in the United States of America.
9 8 7 6 5 4 3 2 1

Table of Contents

Foreword		8
I	Introduction	11
II	Peak Oil	17
III	A Giant Example of Peaking Production	22
IV	Energy Security	25
V	Understanding the Past: A Cartel by Any Other Name	30
VI	Nationalization and the Fall of the Sisters	32
VII	The Myth of Big Oil and the New Cartel	38
VIII	Oil: The Political Weapon	43
IX	Setting the Stage for the Crisis of Today	49
X	The Diesel and Refineries Situation	54
XI	The Thirst for Energy	61
XII	The India Example	64
XIII	The Braking Point	71
XIV	The State of Alternatives	76
XV	The Second Phase of the Hydrocarbon Age	87
XVI	The Courage to Lead	104
Appendix I		113
Appendix II		119
Index		128
Glossary		134
Energy Conversion Factors		161
About the Authors		162

Acknowledgments

I want to dedicate this book to Nancy, my wife of thirty years, and to my three children and their spouses: Joe, Matt and Mandy, Aubrey and Josh; to my parents, George and Lucy Stansberry, and my sister, Mary Fern Carpenter; to all my friends, supporters, and mentors throughout the years. I especially want to thank former U. S. Senator Dewey F. Bartlett, J. Cooper West, Sherman Smith, his family and staff, Mac Alloway, Wendell Powell, Robert Gerry III, Eddie L. Moore, Energy Ventures Group, Jim Nance, W.A. "Bill" Snyder, Dr. Charles Mankin, Roger Beavers, Duke Ligon, Dr. Dennis O'Brien, and all my energy associates and advisors throughout the years.

Also, we would like to thank all of the people that assisted us in writing this work. It has been a labor of love, and years in the making. Special thanks to His Royal Highness Reza Pahlavi and Charley Maxwell for their insightful and thought provoking interviews. Their input has been very important in the creation of this book, and it is much appreciated.

Jason Reimbold would also like to thank the many individuals that have both mentored and challenged him to make a difference in his chosen industry. Thanks go to his wife Tina, their two children, Ariel and Blake, and his extended family for their encouragement and support. Furthermore, Jason would like to thank Michael Bellovich and Rick Batlle for research help and suggestions.

Finally, Jason thanks Fred Boyle, Christyn Champagne, Mickey Coats, R. Dobie Langenkamp, Beth Robinowitz, Jerry Robinson, Dale Steinkuehler, and Dan Woods for their continued friendship and support.

Foreword

I first recognized what a real force OPEC was in 1975 while working for then U. S. Senator Dewey F. Bartlett. Senator Bartlett had asked several of his staff members including myself to review remarks he was going to make in Norway before OPEC officials. Only a year before, the energy industry had been deeply impacted by the 1973-74 oil embargo. It was obvious that the energy industry and our nation's petroleum security would be important issues to deal with during my lifetime.

In 1977, I began working as a petroleum landman. I didn't realize at that time that some of the oil and gas leases I was purchasing would wind up as locations for some of the deepest natural gas wells not only in the Anadarko Basin in western Oklahoma but the world (over 20,000 feet) and that natural gas would be playing a significant role in meeting the energy needs in the 21st Century. However, the boom our industry experienced in the late 1970s and early 1980s attracted the attention of our policy makers, and there was an attempt to implement a Windfall Profits Tax. I remember being quoted in an

Oklahoma City newspaper in 1980 as saying that the Windfall Profits Tax would be devastating to the energy industry. Twenty-six years later, California Proposition 87, a tax proposal on the oil and gas industry, was presented to California voters and fortunately defeated. Nevertheless, this demonstrates the lack of communication between our industry and our government.

Another condition that complicates the global energy balance is the world's growing energy demand and our outdated infrastructure. In the early 1990s, I led a business delegation to China and went to Russia as a member of an energy delegation. In China, we met with government officials regarding natural gas development—especially compressed natural gas. In Russia, we met with the Minister of Energy, and we were told the Russian government had identified 35,000 idle wells in need of repair. Unfortunately, similar conditions still exist today, only now, demand is far greater. This outdated and inefficient infrastructure is making it even more difficult to meet the demand for energy that exists today. During the 1990s, I became very active in providing due diligence in regard to oil and gas mergers and acquisitions in over twenty states and several countries. At that time, there was very little exploration, but by the 21st Century, energy exploration had begun to pick up in a major way.

In April 2004, I served as the keynote speaker of the annual meeting of Merit Energy in Dallas, Texas. There I announced that I was beginning the book entitled *The Braking Point*. With Jason Reimbold's great assistance, the book is in your hands today.

Thirty years ago, I began my journey in the energy industry. Hopefully, we will leave a great energy industry in place for future generations. It is up to us!

— Mark A. Stansberry

I
Introduction

In late 2006, U.S. headlines read:

**Oil Prices Creep Higher on Terror Fears
Nigeria Villagers Seize 3 Oil Platforms
Australia Faces Blackouts**

America, and the world, is facing an energy crisis. The good news is that America's energy problems can be solved. The bad news is that the energy crisis does not exist in a vacuum, it is not temporary, and it will not fix itself. Failure to effectively deal with this problem now will threaten our nation's economic prosperity, compromise our national security, and could radically alter our way of life. The major conditions that dramatically define the energy problem facing America today are: peaking world oil production, energy security, and inadequacies in domestic and international energy policy. This book is a primer on those conditions and is meant to inspire a discussion in this country about the necessary changes which must take place in our industries, consumer patterns, and our government if we are to meet the challenges ahead. Communication between consumers, policy makers, and industry is essential if we are to be successful. The situation is complex however, and there are many factors contributing to what could lead to a time when social and economic progress is halted due to a lack of energy—*the braking point.*

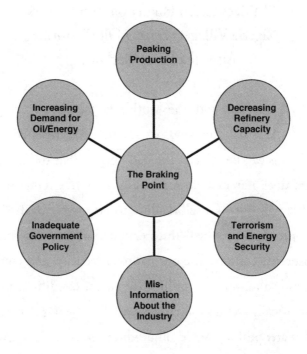

The demand for energy will continue to increase. In the next 20 years, according to estimates by the Energy Information Administration, (EIA), America's demand for oil will increase by 40 percent. As consumption surges, U.S. production will continue to drop precipitously. In 1970, the United States imported 36 percent of our total consumption. As of 2006, the U.S. imported nearly 60 percent of our daily consumption. By 2020, if we fail to meet this challenge, we will have to rely on foreign governments for nearly 70 percent of our oil. This will present the U.S. with even greater security risks than it faces today.

Many of the same issues confront the future of natural gas. America's demand for natural gas is projected to rise even faster than oil. If Department of Energy (DOE) projections are correct, by 2020 Americans will consume 62 percent more natural gas than we do today. The EIA also estimates that approximately 40 percent of the potential gas reserves in the U.S. are located on federal lands that are either closed to exploration or severely restricted. The notion that we can meet the increase in demand for gas, maintain severe restrictions on exploration, and still enjoy low prices is unfounded. This leads to the next problem.

Growth in energy supplies is currently being curtailed by a regulatory structure that has failed to keep pace with advances in technology and often discourages investment in desperately needed facilities. Billions of barrels of oil and trillions of cubic feet of gas lie buried in locations around the United States that are *off limits* to oil and gas exploration due to local, state and federal restrictions. In 1980, for example, Occidental Petroleum discovered an oil field containing several billion barrels of oil in the Catalina Channel just a few miles offshore between Los Angeles and Santa Barbara. To recover this oil, the company had to build a number of Texas Towers within sight of the homes along the coast. Local citizens formed a group called "Get Oil Out," or GOO, which was successful in preventing the recovery of these deposits. In the last 20 years, this same situation has been repeated countless times in different locations across the United States with the same result. Yet these deposits are still in place. Until recently, the American consumer had to choose between environmental reform or unsightly energy production facilities with low energy costs; however, these choices are beginning to disappear. In the next 20 years, we may not have the luxury of choosing a hillside view over the procurement of resources.

Our energy infrastructure — that network of generators, transmission lines, refineries, and pipelines that get raw resources to the consumer in the form of useable fuel — is woefully antiquated and inadequate to meet the future needs of the country. Even if we find enough natural gas to meet growing demand, moving that gas to market will require additional pipelines and distribution lines, and building this network of pipelines will cost a considerable amount.

Additionally, the lack of refining capacity in the United States has a significant impact on the supply and price of gasoline. Since 1980, the number of American oil refineries has dropped by nearly 60 percent (DOE). **There hasn't been a new refinery built in the United States in over 25 years.** New regulatory interpretations limit the ability of existing refineries to expand capacity. Furthermore, current regulations require the production of more than 15 different types of gasoline, and therefore, the refining industry is strained to capacity. These conditions are leaving the United States dangerously vulnerable to regional supply disruptions and price spikes.

Finally, *it is in the interest of national security for the U.S. to further develop its domestic reserves.* As growing economies, namely India and China, seek the energy to realize their full growth potential, global exploration and production (E&P) activities have become highly competitive and politically charged; given the current geopolitical climate and global attitudes towards the United States, it will be necessary to adopt new domestic strategies for procuring energy. Succinctly: The world's demand for energy is rapidly increasing, and geopolitical tensions in hydrocarbon rich areas of the world make the United States economically vulnerable. Today, our energy policy is largely inadequate for dealing with the looming energy crisis.

II
Peak Oil

Oil and gas, and their derivatives, touch nearly every aspect of our lives. Across the globe, cars, irrigation systems, plastics, and factories depend on the production of oil and gas. For over a hundred years, the U.S. has been producing hydrocarbons for energy. Initially oil in the United States was produced for refining kerosene used for illumination. However, after the advent of the internal combustion engine, gasoline had a use, and the world began to take the shape we recognize today.

In our modern life, petroleum-based products are linked to nearly every facet of our lives. Just to name a few, oil and natural gas contribute to the production of:

- gasoline
- plastics
- paint
- fertilizer and
- power generation.

There has been much discussion of the world running out of oil, and fears have been fueled by marked increases in consumption and dwindling new discoveries. These conditions have put oil at the center of the world's political stage which injects another risk factor: energy security. Now, more than ever, governments, economies, businesses, and citizens are wondering when we will run out of oil.

Oil production is a vastly misunderstood process. Oil is not found in great underground caverns or lakes; oil is produced from rock. Much like a sponge that holds water, porous rock holds oil under very high pressures in a given zone. This is called reservoir rock. When a company drills into the rock, the borehole leading to the surface allows the oil to escape. This is oil production. However, this production does not last forever. Throughout the life of a well, the pressure gradually weakens, thus producing fewer and fewer barrels of oil per day.

It is important to clarify the term *reserves* at this point. Oil *reserves* are not an estimate of total oil available in a given section; rather, *reserves* is an economic term describing the amount of oil that can be produced economically from a given reservoir. Therefore, market prices greatly affect reserve estimates, and estimates change frequently. Depletion of reserves does not mean that oil is running out. It simply means that all recoverable oil from that well has been exhausted. Sufficient pressure to produce economic quantities of oil is lost long before the oil is completely gone. In fact, it is estimated that in the United States, oil fields are *depleted* with nearly 67 percent of the oil still in the ground. (DOE)

In 1956, geophysicist Dr. M. King Hubbert proposed that oil production follows a bell-shaped curve; thus, a new discovery brought on an exponential increase in production. However, upon reaching the point at which half of the reserves have been produced, the production declines precipitously; this is the Peak Oil theory. The logic is rather simplistic, but it yields much merit. Certainly, the production of oil wells depletes their reserves; however, secondary and tertiary recovery techniques have skewed Hubert's bell-shaped curve, and many wells produce economic quantities of reserves beyond engineering expectations. Yet, the gradual depletion of reserves is a serious issue when combined with explosive increases in demand.

The International Energy Agency (IEA) projects global demand for oil to increase by 50 percent by 2025. Where is this oil going to come from? Some expect world oil production to peak near this time horizon, and the precedent for this concern was the peaking of U.S.

production in the lower 48 states during the 1970s.

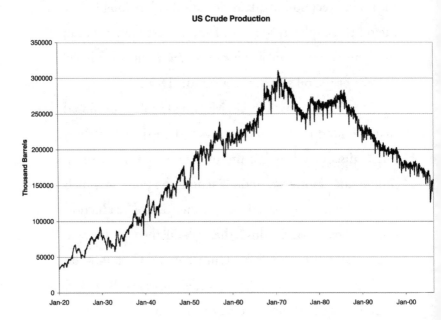

Figure 2—Source: EIA Data

During the 1970s, U.S. consumption began to outstrip production, and America started importing more oil than it produced. **The EIA projects U.S. oil consumption to increase by 28 percent by 2030** (EIA Annual Energy Outlook 2007). However, as consumption increases, oil production in the United States will likely continue its downward trend. Today, the United States imports approximately 60 percent of our total oil consumption (EIA), and if this trend continues, the U.S. will soon rely on foreign governments for more than 70 percent of our oil.

In October of 2006, OPEC began discussing cutting production by more than a million barrels per day, and their intent was to maintain oil prices above $60 per barrel. **Clearly, OPEC does not operate with the best interests of the U.S. in mind.** Furthermore, the peaking of production in the United States is not the primary concern anymore; rather, analysts across the globe are looking for signs of peak production occurring in the Middle East—specifically, Saudi Arabia.

In 2005, investment banker Matt Simmons wrote the book *Twilight in the Desert* in which he discussed the possibilities of the oncoming Saudi oil shock. Unlike the United States, the members of OPEC do not adhere to a standardized form of reporting production figures, and given the strained relationship between OPEC and the rest of the world, some analysts speculate that reserves are overstated, and some even speculate that Saudi Arabian oil production has already peaked! If this is the case, we could be closer to the *braking point* than is realized.

III
A Giant Example of Peaking Production

Much of the concern with Saudi oil production is centered on the world's largest oil field, Ghawar. Located just west of Dhahran in the Kingdom of Saudi Arabia, the Ghawar field stretches over 3000 square kilometers and produces nearly 5 percent of the world's oil each day. Ghawar was discovered in the 1950s by Saudi Aramco, which was owned by BP before the nationalization of the oil company by the Saudi government in 1980. The oil field sustained aggressive development through 1968, and the field is still under development today.

Yet most analysts (including Matt Simmons) are concerned about the method by which Ghawar was developed. Because of the complications that can occur during oil field development, overproduction can lead to a rapid loss of pressure, thus making oil production more difficult and more expensive.

In fact, in 1966, Aramco began a secondary recovery technique called water flooding. This is a method of

A Giant Example of Peaking Production

sustaining reservoir pressure by injecting water into a reservoir using a system of high-pressure pumps. At this point, all sounds well. Water is being injected, pressure is sustained, and oil is being produced. This was the case for several years, but by the end of the 1970s, water was showing up in the oil. Ghawar's oil/water ratio was approximately 70 percent oil and 30 percent water, otherwise referred to as watercut. This began to raise some concern, and because of this, production was cut back dramatically. Despite these attempts to regulate production, the alarms sounded when Ghawar experienced its first water breakthrough in 1983.

In order to address this concern, the Saudis worked over the problem wells, and they were successful in bringing watercuts back to a relatively consistent level. The EIA now estimates Ghawar has approximately *52 percent* of economic reserves remaining in the reservoir; this is nearly the definitive example of peak production. Yet there is a remaining 2 percent left before the world's largest oil field officially peaks, but what's a couple hundredths between friends? With Ghawar's near 28 percent average decline rate, development of other reserves must occur at a frantic pace and must continue to maintain enough production to meet the thirst of the world.

Compounding this problem the Saudi economy is almost solely dependent on the export of light crude oil, specifically called **Arabian Light**. Light or sweet crude requires less refining and is therefore the Grail of oil. Other types of heavier crude are also produced in the Arabian Peninsula. However, this lower quality, higher sulfur oil does not yield the profit of Arabian Light, and **Ghawar is the largest producer of light crude in the Kingdom**. There have never been specifics as to the extent of work performed, and normal production was realized in the 1990s, but again, how confident can we be of Aramco's reporting? Now, with the world looking to the Kingdom of Saudi Arabia and of the Middle East for oil, the balance between their interests and ours is very delicate.

IV
Energy Security

In the last fifty years, the U.S. has looked abroad for other sources of oil, most notably the Middle East. According to the *Oil and Gas Journal* (2005), the Middle East possesses nearly 62 percent of the world's known oil reserves, making it the center of attention in the eyes of the world. Unfortunately, the Middle East has been politically unstable for centuries and many say that political tensions in the Middle East have worsened due to world interests in the region.

Compounding these difficulties, U.S. policy regarding the Middle East has been less than effective. The United States' open support of Israel during the 1973 Arab-Israeli War led to the Arab Oil Embargo against the United States. Being cut off from the world's largest oil reserves produced a shockwave felt around the world. This was America's first real experience with an oil shortage, and though the embargo was brief, it began the discussion of energy security.

America's last three economic recessions have been linked to high energy prices, and there is much speculation that escalating oil prices are having a negative impact on the longest period of prosperity in this country's history. The National Association of Manufacturers estimates that soaring fuel prices between 1999 and 2000 cost the U.S. economy more than $115 billion. **Diminishing oil supplies and rising prices are challenges that threaten the very foundation of our economic livelihood.**

Since 2002, President Bush has discussed the need to alleviate our dependence on foreign oil supplies, and in particularly the need for the development of alternative energies. (*See Appendix I for the White House Energy Bill Press Release.*) Although the President has spoken of hydrogen energy, this proposed solution appears to be feasible only in the very distant future. However, hydrocarbons are here to stay for the near term. Yet, this does not mean policy adjustments cannot be made to improve the U.S. position in the world oil trade. America is fearful of depending on foreign oil, and rightfully so; the EIA recently published the following World Energy Hotspots, and what they discovered is quite alarming.

Energy Security

The following chart shows U.S. imports by source:

Country/Region	Petroleum Prod'n (2004) ('000 bbl/d)	Petroleum Prod'n (2010) ('000 bbl/d)	U.S. Imports (Jan-Mar '05) ('000 bbl/d) 1	Strategic Importance/Threats
Algeria	1,900	2,000	414	Armed militants have confronted gov't forces
Bolivia	40	45*	0	Large reserves of NG (24 Tcf), exports may be delayed due to controversial new laws unfriendly to foreigners.
Caspian Sea	1,800	2,400 - 5,900	0	BTC opened, many ethnic conflicts, high expectations for future oil production, no maritime border agt.
Caucasus Region 2	Negligible	Negligible	0	Strategic transit area for NG and oil pipelines
Colombia	551	450*	110	Destabilizing force in S. America, oil exports subject to attack by protesters, armed militants.
Ecuador	535	850*	316	Unstable politically, protests threaten oil export
Indonesia	900	1,500	0	No longer a net exporter, separatist movements, Peacekeeping forces in place, Violence threat to Strait of Malacca
Iran	4,100	4,000	0	Even though no direct imports to U.S., still exports 2.5 million bbl/d to world markets
Iraq	2,025	3,700	516	April 2003-May 2005 - 236 attacks on Iraqi Infrastructure
Libya	1,600	2,000	32	Newly restored diplomatic relations, Western IOCs not awarded contracts in 2nd EPSA round
Nigeria	2,500	2,600	1,071	High rate of violent crime, large income disparity, tribal/ethnic conflict and protests have repeatedly suspended oil exports
Russia	9,300	11,100	419	2nd only to S.A. in oil production, Yukos affair has bred uncertain investment climate
Saudi Arabia	10,400	13,200	1,614	Long term stability of al-Saud family, Western oil workers subject to attacks
Sudan	344	530*	0	Darfur crisis & N-S conflict threatens government stability, security of oil transport
Venezuela	2,900	3,700	1,579	Large exporter to U.S., President Chavez frequently threatens to divert those exports, nationalize resource base

*Estimate or non-EIA source
[1] Although a country may not have direct exports to the U.S. it may still have significant exports to world markets
[2] Includes production from Georgia and Armenia. Azerbaijan production included in 'Caspian'.

Figure 3—Source: EIA Data

The total crude oil imported from identified threats for this time period was 6,071Mbbl/d; this is nearly 51 percent of total U.S. imported oil. These seem to be very high percentages given the fact that our relationship with many of these nations is strained at best. Most of these nations are anti-American!

Some speculate anti-American sentiment is largely the result of our involvement in Iraq, but ill will towards the land of liberty has plagued us and the region for quite some time. The Middle East was unstable long before either Gulf War largely as a result of social oppression and economic depression brought on by inequities of monarchies, government-owned industries, and widespread corruption. The rulers have become very wealthy while the *ruled* are impoverished. This gap is what has most analysts, corporations, small businesses, and consumers very anxious about the petroleum security of the United States because it leads to further political instability. Yet despite this issue, the American public largely blames Big Oil for the recent increase in oil prices. This idea of the J.R. Ewings profiting at the expense of American consumers is grossly misguided. For example, there are thousands of individuals that operate marginal wells (wells that produce 10 barrels per day or less). This marginal production not only provides a living for the producers and their families, but it also accounts for nearly 60 percent of our domestic production. It is true that during the infancy of the oil industry, corruption and anti-consumer practices—such as supply manipulation and price fixing—were prevalent. There have always been rogue elements of corruption in free markets, but the days of industry-wide corruption and conspiracy are not nearly as commonplace as in the past.

V
Understanding the Past: A Cartel by Any Other Name

In the oil industry, cartels have been common. The purpose of a cartel is to protect profits through uniform guidelines and actions. Members of the group can do this by expanding or contracting the supply of a commodity depending on the demand conditions at any given time. The first famous cartel in the oil industry came in the 1920s and lasted until the early 1970s. This cartel dissolved only to be replaced by another.

Following the U.S. government-mandated breakup of the Standard Oil Company, a cartel composed of seven multinational petroleum firms came into existence. The companies involved were British Petroleum, Royal Dutch Petroleum/Shell Oil, Socony-Vacuum (now part of Exxon-Mobil), Standard Oil of California (now Chevron), Standard Oil of New Jersey (now part of Exxon-Mobil), Gulf Oil and Texaco. This group of companies was referred to as the "Seven

Sisters," a term originating in the mind of Italian oil entrepreneur and ENI Chairman Enrico Mattei. Mattei originally used this term in a derogatory fashion when referring to the stranglehold that these firms had on the petroleum industry. He also felt the need to speak out against the Seven Sisters because as the chairman of the National Oil Company of Italy, he was protecting the interest of his firm and of himself. The name was also used as a public commentary against the cartel-like behavior of the Seven Sisters.

VI
Nationalization and the Fall of the Sisters

The Seven Sisters were instrumental in controlling the price of oil. They possessed this power for the greater part of the 20th century. Their alliance can be traced back to a pact signed between the companies on September 17, 1920. During their reign, the Sisters had control of roughly 90 percent of the world's oil and earned profits of 85 percent while oil-producing countries earned the residual profits. These firms were able to maintain high control of the world's oil-producing, exploring, refining, and transporting ability through the cooperation of many joint ventures. These companies worked closely together in order to maintain power. They also preserved their power because the oil-producing countries did not have the resources necessary to fight the multinational oil firms. These firms had a competitive advantage because they were easily able to explore, refine, and ship oil much more effeciently than any of the nations where the oil existed. These firms operated in developing nations that were not able to fight the machine that was the Seven Sisters. In addition to the Sisters' vast operating power, the companies had the backing of many of the United States' most powerful banks.

Through this cartel structure, the oil companies and the banks profited greatly from working deals together. With each company backing each other, and with the help of many powerful financial institutions, the Seven Sisters seemed unstoppable.

However, the first sign of a changing of the guard came in 1938 when Mexican President Lazaro Cardenas nationalized the nation's oil. Cardenas made this move to prevent further Americanization of Mexico. This move pushed all of the American oil companies out of Mexico. One of Cardenas' motives to nationalize the Mexican oil industry was to cement the President's popularity on the heels of the Mexican Revolution. This move also gave Mexico full control over the production and export of oil. In response to this nationalization of Mexico's petroleum industry, the U.S. boycotted Mexican oil to keep other countries from attempting to implement a state-owned oil industry model.

The effect of nationalizing Mexico's oil industry was

mixed. The country received very little compensation from the U.S. through the settlement between U.S. oil companies and the Mexican government. This move also ensured that Mexico would not receive any forms of foreign investment and the government-owned oil companies did not have the core competencies necessary to run successful oil firms. Mexico could not produce enough oil to meet even the demand of its own people. By 1957, Mexico imported more oil than it could produce in order to meet the domestic demand. To this day, the country imports more oil than it produces. This reality ensured that the Seven Sisters would still be able to operate.

Although the nationalization of Mexican petroleum did not help the people of Mexico and did not topple the vast power of the Seven Sisters, the move set the tone for things to come. In 1951, Iranian Prime Minister Mohammed Mossadegh pushed for the nationalization of the nation's oil industry. This idea, which was well received by the country's parliament, came after it became publicly known that the Anglo-Persian Oil Company was receiving 85 percent of the revenues from production of Iran's oil. Iran's parliament easily passed a bill that would nationalize the oil industry. In order to prevent this move, the British government offered to share half of its profits with the Iranian government, but Prime Minister Mossadegh considered the offer unsatisfactory. This move resulted in immediate disaster for the Iranian oil industry. Many nations disagreed with Iran's imposed embargoes. The petroleum industry in Iran collapsed, and Mossadegh was removed from power in 1953. Soon after these events, the British Petroleum Company (formerly the Anglo-Persian Company) signed a new agreement with Iran and was allowed to operate within the country again.

In 1943, Venezuela signed an agreement that would impact the Seven Sisters greatly. The government changed its tax structure so that oil companies who wanted to operate in the country would be required to provide a predetermined amount of money to the government in addition to splitting half of the company's profits with Venezuela. This was known as the "Fifty/Fifty Principle." By the latter part of the 1940s, Venezuela would demand more money from oil producers by changing its tax system further. In return, the oil companies moved from Venezuela to countries with more favorable conditions. Venezuela soon contacted other oil-producing countries and urged them to demand higher shares of petroleum profits.

The effect of the 50/50 agreements implemented by many countries hurt the Seven Sisters greatly. By the 1950s, world oil demand grew rapidly, but the existing supply was much higher. In simplified macroeconomic terms, the supply of oil far exceeded the demand. Therefore, the price of oil was low and the oil companies generated lower revenues. These agreements were based on price quotas and not the market value of petroleum. This aspect of the market caused oil companies to generate less profits. Not only did these firms receive less profits, but they were also paying more in royalties to the oil-producing nations. The response to this problem came in the form of artificial stimulation of the market. The U.S. government was the first to implement a new oil policy. The U.S. was moving from a net oil exporter to an importer, and President Dwight D. Eisenhower felt that this greatly compromised national security. If the U.S. had to rely on others for oil, then the nation might have to agree to another country's terms both politically and economically. In order to gain better control of the market, Eisenhower enacted an import quota. This allowed the price of oil to rise sharply and benefited U.S. oil companies. The multinational firms cut their posted prices in 1958 to secure better market share and profits. British Petroleum also cut its prices. This led to more money for the oil companies and once again oil producing countries received a smaller share of the profits. These actions were primarily responsible for a drastic changing of the guard in the petroleum industry.

VII
The Myth of Big Oil and the New Cartel

No more are the days of the oil barons and industrial titans. Big business has become public business just as big oil has become public oil. Shareholders own the super-major oil companies. Their profits are the shareholder profits. Yet *big oil* is still suspect in the minds of consumers. It is true the Seven Sisters fixed prices at one time, but with the advent of OPEC, those days have long since vanished.

Drastic cuts in the price of oil in 1958 and 1959 led to the formation of a new cartel. On September 10-14, 1960, several oil producing companies met in Baghdad to discuss the current state of the oil industry and the role of oil-producing companies in the market. The result of these meetings was the Organization of Petroleum Exporting Countries, formed to "coordinate and unify petroleum policies among member countries, in order to secure fair and stable prices for petroleum producers; an efficient, economic and regular supply of petroleum to consuming nations; and a fair return on capital to those investing in the industry." (IMF Directory of Economic, Commodity and Development Organizations) The main function and intention of this group was publicly stated. The organization set out to form a cartel in order to secure profits in the oil industry by controlling the supply of oil in the world through agreed-upon production quotas. The main reason that OPEC was created was to respond to the Seven Sisters' desire to control the price of oil. Many members of OPEC believed that these firms wanted to lower prices in order to pay lower royalty amounts to the oil-producing countries. The founding members of OPEC were:

- Iran
- Iraq
- Kuwait
- Saudi Arabia
- Venezuela

The beginnings of OPEC were humble. Their main goal was to fight the exploitative power of the Seven Sisters. At first, OPEC had very little power. In its first decade of existence, the group mainly charted out its policies and guidelines and remained out of the public eye. During the 1960s, OPEC was able to stabilize oil prices, but the group could not reap a greater amount of profits. The group only controlled 26 percent of oil interests worldwide.

In order to command more money, the oil-producing countries would have to take control of mineral rights. The multinational firms still owned the rights to underground oil deposits in these nations. To gain ownership of this oil, the Middle Eastern countries would have to look past each respective political difference and push to nationalize operations in every oil producing country. These countries were also still developing and by running off the multinational petroleum firms, these economies would no longer benefit from these oil operations. OPEC remained limited in power until the 1970s.

On September 1, 1969, Muammar al-Gaddafi rose to power in Libya. Soon after he took office, Gaddafi demanded higher royalties from the oil companies operating in the nation. He ordered these firms to increase royalties from oil production by roughly 20 percent. This led to the first 55/45 agreement, which caused a chain reaction as many other nations required similar conditions be met in order to operate. This meant greater control of oil by the producing nations and shifted control from the multinational firms. The first prevalent public media exposure for OPEC came from policies implemented in response to the Yom Kippur War. This war resulted from years of Arab-Israeli tensions and many believed that any disruption of peace in that region would cause an energy crisis. Many people, especially in the United States, enjoyed low oil prices that resulted from supplies that were greater than demand. Any disruption in that routine would cause great discomfort for the people around the world. In fact, the economies of several countries including the U.S. and Japan depended greatly on reasonably low energy prices—but disruption was soon to follow! Enter the Petro-State.

With the newfound wealth of controlling their own production, the nations of OPEC began to enter the global political stage, and though their oil was needed, their attempt to participate economically with the rest of the world was greeted by a cool reception. Other nations around the world, including the United States, held the producing nations in low regard. Unfortunately, the consuming nations made one grave economic miscalculation. If OPEC restricted supply, they would make even more money while simultaneously throwing the consuming nations into economic chaos. In this way, Saudi Arabia became the swing producer of the cartel; that is to say that because Saudi reserves were so immense, they could manipulate supply all on their own, and thereby set the price for the world's oil. Already at this time, the United States was largely dependent on the oil from the Middle East, a region where we had few friends.

VIII
Oil: The Political Weapon

Throughout industrialization, the United States thrived on cheap oil. In the years following the end of World War II, the U.S. would double its oil consumption and enjoy the economic development fueled by affordable oil. However, this prosperous stretch would begin to fade in the wake of increasing inflation in the late 1960s and early 1970s. At that time, the U.S. dollar was the universal currency of trade, and more importantly, it was the currency in which oil was traded. In order to support the stagnant economy, President Richard Nixon implemented a counter-intuitive strategy. Nixon would end the *Bretton Woods* system and take the United States off of the gold standard. This move shocked the world economy, and the dollar's value began to plummet. This of course seemed like a terrible strategy until oil entered the picture. Economies benefit from inexpensive oil, and the oil of OPEC—specifically Saudi oil—was priced and sold in U.S. dollars. The deflation of the dollar value forced the price of oil down. This sent a shock through the Middle Eastern producing nations as the value of their primary — and in some cases *only* — product diminished. **Not only did this action strain** the United States' relations in the region, but other geopolitical events were about to take place that would lead to the flexing of the Middle Eastern oil muscle.

At this time, the United States was not directly an adversary of the Middle East. In fact, the U.S. enjoyed a very positive relationship with Iran — formerly Persia — for many years. Much of the anti-Western sentiment in the late 1950s and early 1960s was directed at European nations, namely England, for the exploitation of Mid-East oil. It was after the *Six Days War* of 1967 that the Arabic members of OPEC formed a sub-organization. The **Organization of Arab Petroleum Exporting Countries (OAPEC)** was established for the express reason of exerting pressure on the western nations pledging support to Israel. In time, the non-producing nations of Egypt and Syria also joined OAPEC; this was the beginning of an organized and official political movement against the West. This is not to say that the United States was held in high regard, just that we were not the focus of ill will, at least not until the Yom Kippur War of 1973. This offensive was a collaborative effort amongst the Arab nations to reclaim land from Israel. However, this would be the first time the Arab (OPEC) nations would wield the oil weapon. At the onset of the war, OAPEC, including Syria and Egypt, implemented an embargo against the Netherlands and the United States because of their support of Israel.

At a time when America was experiencing upheaval and uncertainty on nearly all socio-economic fronts, the energy crisis threatened the very fabric of the American way of life. This was addressed in Nixon's speech to America about the energy crisis.

"America's energy demands have grown so rapidly that they now outstrip our energy supplies. As a result, we face the possibility of temporary fuel shortages and some increases in fuel prices in America.

"This is a serious challenge, but we have the ability to meet it. If our energy resources are properly developed, they can fulfill our energy requirement for centuries to come.

"What is needed now is decisive and responsible action to increase our energy supplies, action which takes into account the needs of our economy, of our environment, and of our national security, and that is why I am moving forward today on several fronts.

"I am ending quantitative controls on oil imports and establishing a National Energy Office. I am ordering an acceleration in the leasing of oil lands on the Outer Continental Shelf and increasing our ability to prevent oil spills.

"I am also taking new steps to maintain our vital

coal industry. In addition, I am asking the Congress to act quickly on several proposals. One would remove government regulations which now discourage the growth of our domestic natural gas industry. Another would help us establish the research and technological groundwork for developing new forms of energy with a long-range future. And still others would permit licensing of new deepwater ports in our oceans and would open the way for the long delayed Alaska oil pipeline.

"*Each of these steps can help us meet our energy needs and meet those needs without sacrificing our environment.*"

The 1973-74 embargo sent a shock through the world; the OAPEC nations cut oil production and embargoed shipments of crude to the United States. Prices reacted almost instantly, and the effects of the Arab oil weapon were felt by citizens across America. Though President Nixon demonstrated courage by abolishing the gold standard, he unfortunately did not exercise such courage in battling the embargo. Rather than taking the political fight toe to toe as he did previously when the price of oil plagued the U.S. economy, this time Nixon implemented price controls; this only intensified the problem, and American consumers suffered. By artificially keeping oil prices low, shortages soon followed because demand destruction was not able to occur. Now,

America experienced the magnified effects of being held hostage of OAPEC. Nixon understood the problem, but not the war. From May to June 1973, the price of gasoline in the U.S. increased 43 percent to $.55, and in six weeks, the New York Stock Exchange lost $97 billion in value. Fuel shortages were commonplace during the period of September 1973 to summer of 1974. This was the first time since World War II that U.S. citizens experienced lines at the pump. During this period, daily consumption in the United States decreased by *7 percent*, but **Arab oil imports were almost all but diminished, dropping 98 percent**. This cut in the global supply of oil was devastating to the U.S. economy, and soon other western nations experienced high inflation and economic recession.

To help alleviate these effects, President Nixon implemented price controls and rationing methods. However, as is often the case with government intrusion on free markets, these acts largely failed. Some of the methods to ration gasoline and curb consumption included the following:

- Cars with license plates ending in odd/even numbers were allowed to fuel on alternate days.

- The national speed limit was reduced to 55mph.
- Year-round daylight savings time was implemented.
- Requirements for fuel efficiency standards in American made automobiles were implemented.

Despite these measures, rather than because of them, the effects of the embargo began to soften as the price of oil began to fall. OAPEC failed to consider the long-term effects of high oil prices, including the changing of exploration economics which made offshore discoveries like that of the North Sea possible. Further changes brought on by the high price of oil included expansion in alternate energy sources such as the following:

- nuclear generation of electricity
- ethanol blending of gasoline stock, and
- wider use of natural gas for power generation and home heating.

In 1973, we saw the first impact of the oil weapon, and President Nixon recognized the importance of national petroleum security. Today, the price of oil is primarily set by free markets. Although, pricing can be manipulated by OPEC. (*For a chronology of oil prices, see Appendix II.*)

IX
SETTING THE STAGE FOR THE CRISIS OF TODAY

In 1979, the United States would be engaged yet in another energy crisis. This time, however, America's political policies were not the cause of the problem, but America would still be the victim of attack. During the time of high oil prices, the oil producing nations were awash in revenues. Unfortunately, these revenues did little to improve the Middle East's economic infrastructure. Furthermore, religious clerics were gaining political power. Perhaps the greatest example of this was the revolution in Iran and the removal of the Shah. Mohammad Reza Pahlavi fled Iran in 1979, and Iran was taken over by the Ayatollah Khomeini. It has been suggested that the United States allowed this transfer of power to happen because the United States was worried about the advancement of communism through the region. The Soviet Union was already making headway into Afghanistan, and the American government believed theocratic governments would be more robust to communist invasion than secular nations. However, the political revolution which occurred in Iran spurred panic throughout the world, and the price of oil began to shoot up once again.

This time, President Jimmy Carter was at the helm, and though he attempted to address the American energy crisis, he had trouble meeting the challenge. In response to this latest crisis, the president delivered the now infamous "Malaise Speech" in which he, in part, blamed American apathy for our energy concerns. Additionally, he outlined what was to be a rather unrealistic plan for addressing the country's long-term energy security.

"***Point one:*** *I am tonight setting a clear goal for the energy policy of the United States. Beginning this moment, this nation will never use more foreign oil than we did in 1977 — never. From now on, every new addition to our demand for energy will be met from our own production and our own conservation. The generation-long growth in our dependence on foreign oil will be stopped dead in its tracks right now and then reversed as we move through the 1980s, for I am tonight setting the further goal of cutting our dependence on foreign oil by one-half by the end of the next decade -- a saving of over 4-1/2 million barrels of imported oil per day.*

"***Point two:*** *To ensure that we meet these targets, I will use my presidential authority to set import quotas. I'm* *announcing tonight that for 1979 and 1980, I will forbid the entry into this country of one drop of foreign oil more than these goals allow. These quotas will ensure a reduction in imports even below the ambitious levels we set at the recent Tokyo summit.*

*"**Point three:** To give us energy security, I am asking for the most massive peacetime commitment of funds and resources in our nation's history to develop America's own alternative sources of fuel — from coal, from oil shale, from plant products for gasohol, from unconventional gas, from the sun.*

*"**Point four:** I'm asking Congress to mandate, to require as a matter of law, that our nation's utility companies cut their massive use of oil by 50 percent within the next decade and switch to other fuels, especially coal, our most abundant energy source.*

*"**Point five:** To make absolutely certain that nothing stands in the way of achieving these goals; I will urge Congress to create an energy mobilization board which, like the War Production Board in World War II, will have the responsibility and authority to cut through the red tape, the delays, and the endless roadblocks to completing key energy projects.*

"We will protect our environment. But when this nation critically needs a refinery or a pipeline, we will build it.

*"**Point six:** I'm proposing a bold conservation program*

to involve every state, county, and city and every average American in our energy battle. This effort will permit you to build conservation into your homes and your lives at a cost you can afford."

Unfortunately, the lack of realistic solutions led to continued strain on the energy policy of America. However, we can credit the President with acknowledging the energy crisis that was thrust upon the United States. Petroleum security was now on the minds of all Americans, and we had our second taste of an energy shortage. *The Prize*, written by Daniel Yergin, discussed how fear influenced people to hoard gasoline. It seems logical to suggest the shortage was mostly due the psychological effect. If everyone in America went to the gas station tomorrow morning to fill up, there would be a gasoline shortage tomorrow morning. The relatively narrow spare capacity of the gasoline supply in the United States cannot accommodate the unscheduled, unlimited refueling of every car in the country. Nevertheless, the energy crisis of 1979 was real regardless of the factors contributing to the shortage.

Compounding the crisis was the *environmental movement* in the United States which was largely prohibitive to both expanding exploration and production in America, and it crippled the refining industry. Therefore, the country was unable to further the production of our own resources, and we could not add to our refining capacity.

In the early 1970s, the Alaska Pipeline was pushed through the regulatory system over the Sierra Club and other conservationists' objections to the effect the pipeline would have on the mating habits of the local caribou. The herd in the Prudhoe Bay area grew more than nine-fold over the past 20 years to an estimated 28,000 animals in 2000 — seemingly irrefutable evidence that caribou and oil exploration can peacefully coexist.

There can be a balance between energy exploration and environmental protection. The marriage of oil and gas exploration with cutting-edge technology means more energy efficiency and enviromental preservation.

X
The Diesel and Refineries Situation

Since the early 1980s, operating refineries have decreased over 56 percent during the period. While some would argue the decrease in refineries has been due to technological advances in equipment and refining technique, the chart below illustrates underlying concern over available refining capacity.

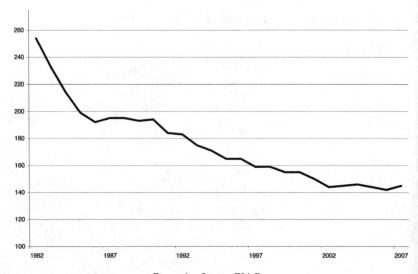

Figure 4—Source: EIA Data

The Diesel and Refineries Situation

According to industry leaders, the efficiencies created in the last few years are creating the equivalent of a new medium sized refinery every year; still, the underlying truth is that even with these efficiencies the demand for petroleum products is outpacing the performance and creation of efficient means of production with no foreseeable end to this trend.

A primary reason that a crude oil refinery has not been opened since 1976 is environmental requirements and permitting. U. S. refineries are operating at record-high levels (up to 98 percent of capacity) to produce the fuels that consumers need. (DOE 2004) Many companies are exporting the refining capacity and storage elsewhere and shipping already refined petroleum productions to the United States.

The cost of gasoline continues to increase, and this is primarily due to refiners paying higher prices for crude oil. Higher tanker freight rates, low European inventories, and more stringent U. S. fuel specifications have contributed to reduced gasoline imports. In fact, here is a listing of the major supply and demand factors that contribute to gasoline costs: (Source DOE 2004)

- High demand
- Low inventories
- Supply snags
- Economic disruption
- Cartel crunch
- The fear factor
- Vulnerable infrastructure

As the list indicates, there are several factors that can contribute to fluctuations in the price of gasoline and other important petroleum products.

Finally, U.S. fuel specifications differ in each state. This causes additional burden on the refiners to produce and create synthetic blends beyond the premium and regular octane gas ratings. Some refineries do not have the capabilities to comply with these standards creating additional transportation costs from outside refineries. Also the laws continue to change requiring massive environmental investments from refiners.

The Strategic Petroleum Reserve has a total capacity of 727 million barrels of crude oil with approximately 60 days of inventory in the reserve. (DOE) There is a maximum drawdown capacity of 4.4 million barrels per day which takes thirteen days to reach the market.

There is no question that America remains

vulnerable as long as we allow imports of foreign oil to continue to increase. America needs to advance new and environmentally friendly technologies, to increase supplies and encourage cleaner, more efficient energy use. **America must update its outdated network of power plants, transmission lines, pipelines, and refineries. Oil pipelines and refining capacity are in need of repair and expansion. Other areas of concern are natural gas distribution which is aging and in some cases inadequate, and an inadequate electric transmission grid. Because of such conditions blackouts and brownouts are inevitable.** An example of aging pipelines was in Highlands Ranch, Colorado, when a gas line burst and threatened the lives and homes of 61 families. As stated earlier in this book, the conditions of increasing consumption, decreasing/peaking production, and reliance on imported oil are bringing the United States, and the world, to the **braking point**!

Major conditions do exist:

- Global demand for energy is rapidly increasing
- Inadequate energy policy is impeding the continued development of energy supplies
- Our energy infrastructure is inadequate

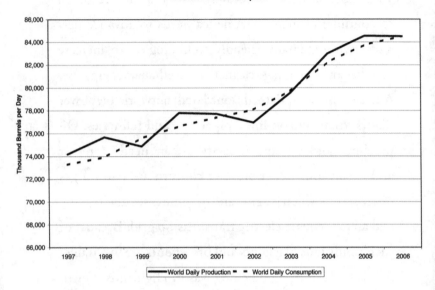

Figure 5—Source: EIA Data

What is needed?

- Manpower
- Research and development
- Technological advancement
- Governmental lands opened up for exploration
- Expansion of refining
- Transportation distribution and
- Storage of natural gas, petroleum products and electricity.

The Diesel and Refineries Situation

Americans comprise 5 percent of the world's population, and we use over 25 percent of the 83 million plus barrels per day of oil processed and a proximate percentage of clean, more abundant North American natural gas. The U. S. Department of Energy estimates that over the next twenty years, U. S. oil consumption will rise by 33 percent, natural gas consumption 60 percent and demand for electricity 45 percent.

The great myth about the energy crisis is that current problems are due to an energy industry that is engaged in a massive conspiracy to gouge consumers by limiting supply to drive up prices. As discussed earlier in this book, there is no magic source of supply; no hidden pool of oil, gas, or electricity that can be turned on and off like a faucet.

California and other power-strapped states will never solve their power crisis until they resolve the conflict between demand and supply. In 2004, a company proposed building a $400 million power plant in California that would have provided enough additional electricity to light 600,000 homes in energy-starved Silicon Valley. The company pledged to plant 800 new trees to beautify the area. They proposed cloaking the power station in a brick facade to make it essentially indistinguishable from a high rent office complex. They even promised to maintain the local habitat for the endangered Bay Checkerspoon Butterfly.

Meanwhile, plans in Los Angeles to build a 550 megawatt gas fired generator were scrapped after residents voted against the project. **Is it really any mystery why there hasn't been a single power plant built in California in so long?**

XI
THE THIRST FOR ENERGY

As the world's consumption continues to increase, it will be difficult for current levels of production to satisfy demand. The International Energy Agency (IEA) recently published estimates of world energy demand increasing by 60 percent by as early as 2030. As these nations continue to grow economically, their consumption of energy should have a corresponding increase. In particular, the GDP Per Capita of the developing nations of China and India has yet to reach the relatively low GDP Per Capita of Mexico.

Country	GDP Per Capita
U.S.	$38,000
Mexico	$9,000
China	$5,000
India	$2,900

Figure 6—Country Per Capita Source: CIA Factbook

Increasing demand is being driven by China, India, and other highly populated developing nations. This rising GDP Per Capita sets the stage for vast increases in demand for energy. During the 2000 to 2005, both India and China increased their energy consumption by 15 percent and 42 percent respectively.

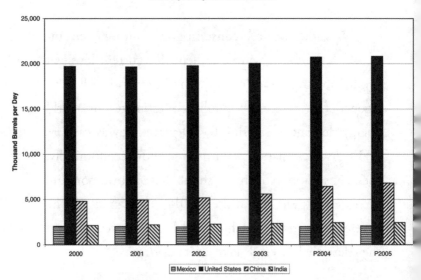

Figure 7—Source: EIA Data

There are approximately 30 million cars in China, but over 1.3 billion people. As globalization continues to evolve and provide the people of China new opportunities, their purchasing power and capacity for consumption will continue to increase. China is expected to be as big a car market as the U.S. by 2015 and could surpass us in capacity by 2035. This adds considerable fuel to the fire that pressure on reserves and further exploration needs to be achieved to be able to satisfy this growing giant.

As the demand for energy increases, oil will continue to dominate the energy mix. Considering IEA estimates, it is important to note that while wind, geothermal, and additional alternative renewable energies show a small increase, the lag time for the implementation of these alternatives will hinder their effectiveness to help meet energy demands in the near and intermediate future. However, capital investments will not be the only change needed to address the issue of increasing demand; **political policy must also adapt to ensure we meet our future energy needs.** An example of this can be seen in the policy changes which have occurred in India as they have attempted to fuel their explosive economic growth.

XII
The India Example

With the fourth largest economy in the world, India is still relatively a new frontier for oil and gas exploration. Until recently, the Indian oil and gas industry has suffered from sizable underinvestment, and this has led to the sub-continent's frenzied quest for energy. Despite significant governmental policy changes since the late 1990s, India has been unable to attract E&P activity from the super-major international oil companies. Sizable offshore opportunities across the globe have been the focus of the super-majors for some time, and this has made India's onshore possibilities appear as an inequitable pursuit – this translates into immense opportunities for smaller independents.

Since the implementation of the New Exploration Licensing Program (NELP) numerous significant onshore discoveries have been made including the potential one-billion barrel discovery made by a joint venture of Cairn Energy and India's largest national oil company ONGC Ltd. in the Managa field. Progressive governmental policies should promote an investment friendly environment for oil and gas ventures.

India's economic growth and expanding need for natural resources have encouraged changes in past policies that limited the sub-continent's exploration for oil and gas. Specifically, the Government of India (GOI) has taken progressive action to attract foreign investment and to soften the budgetary restrictions E&P companies faced in the 1990s. These proactive policies are demonstrated by the recent expansion of oil and gas licensing from an estimated 22 blocks in the 1990s to 94 blocks since 2000, covering an estimated 900,000 square kilometers. At present, India is the world's largest democracy and fourth largest economy; this economic growth has heightened the need for further oil and gas exploration.

GDP Per Capita is a strong indicator of total energy consumption as demonstrated by the model, page 66. It will be important for further development of the Indian oil and gas industry to support the continued expansion of the sub-continent's dynamic economy. In fact, India has had a 51.8 percent increase in total energy consumed per GDP Per Capita.

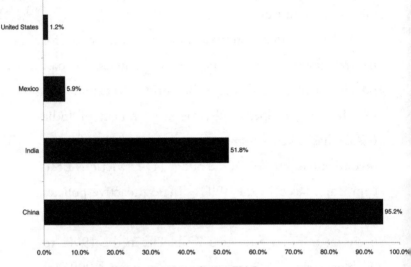

Figure 8—Source: EIA Data

A recent India Brand Equity Foundation (IBEF) report noted that the demand for oil in India has long been in excess of their domestic production. By 2010, India will be the world's fourth largest consumer of oil and natural gas products. These goals have been formalized in the government's *Hydrocarbon Vision 2025*.

IBEF projects India to be the fourth largest energy consumer by 2010. The GOI established its Hydrocarbon Vision 2025 to address the issues of India's present and future energy needs. The primary objectives of the Hydrocarbon Vision included achieving energy security through the environmentally sound development of their domestic resources while maintaining fair competition among public, private, and international companies.

To support the development of India's oil and gas industry, the Ministry of Petroleum and Natural Gas established the regulatory body of the Directorate General of Hydrocarbons (DGH) in the mid 1990s. The primary task of the DGH is to facilitate the effective management of India's oil and gas resources with regard to the economic, technological, and environmental considerations.

These changes in policy have led to an opportunistic environment for investment in India's oil and gas industry. One approach to meeting future demand and achieving energy security is further exploration and development of India's sedimentary basins. Despite the best efforts of the DGH, they report approximately 49 percent of Indian basins have been poorly explored or entirely unexplored.

Proven commercial productivity exists in only 25 percent of the Indian sedimentary basins. Onshore, only 14 percent of the prospective areas have been explored. This is a formidable obstacle to overcome in the pursuit of the goals outlined in Hydrocarbon Vision 2025. Even with the completion of proposed surveys, further corroborating testing would be valuable for assessing the true geologic properties of these prospective basins. In response to these challenges, India's largest National Oil Company (NOC), ONGC Ltd. (Oil and Natural Gas Company Ltd.), and its exploration subsidiary, ONGC Videsh Ltd., are leading the industry in the quest of oil and gas both in India and worldwide.

The exploration of energy sources will concentrate on three major areas. The first is local exploration with an emphasis on offshore investigation, with estimated costs for exploration of $900,000 per day. Second, there are plans for the strategic acquisition of foreign energy assets such as in West Africa, CIS Countries, and in Venezuela and Brazil in Latin America to acquire assets or equity stakes. Furthermore, ONGC will look to exploration opportunities in unstable countries such as in a Sakhalin oil field in Russia and a large oil field in Angola. Moreover, the company also has stakes in oil fields in Iran, Iraq, Syria, Myanmar, and Libya.

While ONGC controls 57 percent of all contracts for oil and gas exploration, the other 43 percent led by Reliance Industries are being filled by private companies. The primary vehicle for the pursuit of energy within the sub-continent has been the efforts of the DGH New Exploration and Licensing Program (NELP).

In 1997, the GOI adjusted its policy of the production sharing contracts (PSCs) in regard to the acreage held by the NOCs; their intent was to level the playing field and thereby attract the foreign and private E&P investment. This change in policy resulted in the NELP program to facilitate the fair competition for exploration acreage.

These changes in PSC terms have led to significant increases in activity. Since the inception of NELP, nearly 100 PSCs have been signed; this is more than 3 times the 28 PSCs that were signed prior to NELP rounds. Additionally, approval time has been reduced to 3-4 month periods, as opposed to 1-2 year time periods of the past.

In addition to policy changes, favorable market conditions have increased opportunity and competition for Indian acreage. Furthermore, exploration activity is markedly higher, and the discovery/wells drilled ratio is improving; thus, the Indian oil and gas industry is demonstrating greater efficiency. Due to the size of these offerings, NELP has not attracted bids from the major international oil companies. At present, NELP participants have primarily consisted of the Indian NOCs; however, several independents are beginning to take interest in the NELP offerings.

The recent policy changes of the GOI provide significant investment opportunities in onshore, upstream ventures. Both the long-term plan of Hydrocarbon

Vision 2025 and the New Licensing Exploration Program provide the foundation for India's surging oil and gas industry. This is a stunning example of how the proper policies can encourage the needed development of a nation's resources.

XIII
The Braking Point

Much of the world's oil comes from fields that were discovered prior to 1970. The addition of new reserves has been decreasing since the 1960s while demand has increased. The graph on page 72, shows how small our margin for spare capacity has been from 1963 to 2002 while world production has increased 50 percent. With the exception of offshore activity, the recent methods of adding to the reserve base have amounted to the exploitation of established reserve areas.

This premise carries a significant risk as many Middle East oilfields are mature, and the reserve additions have dropped to a nominal amount over the last 40 years. Adding to the risk is low excess production. In fact, spare production has only been above 2 percent once in nearly 9 years.

A Goldman Sachs report of March 2005 suggested that, like the conditions of the United States in the 1970s, the further economic development of China would lead

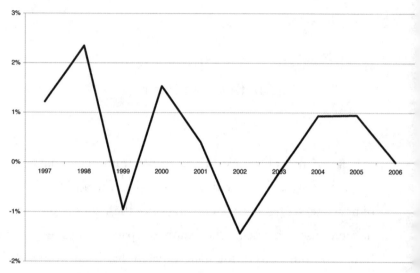

Figure 9—Source: EIA

suggests that developing nations such as China and India will increase the world demand for oil in the future; specifically, the IEA estimates China's demand for oil will increase 3.4 percent per year (although, it seems this is a conservative estimate relative to recent growth performance). Based on IEA projections, significant discoveries of oil will be necessary to meet future demand.

However, not only the analysts have been contemplating the energy crisis. The media in America have begun to take notice as well. Recently, we have seen television productions such as *CNN Presents: We Were Warned*. This program, released in 2006, was a docudrama highlighting the events occurring around the nation during an oil shortage. Another docudrama, *Oil Storm*, produced by the FX Network, aired in 2005. This made-for-TV movie was filmed as a documentary following the riots and panic which spread throughout the United States after a hurricane devastated American oil production in the Gulf of Mexico. Ironically, this movie premiered months before our country experienced Hurricane Katrina. Another example is the feature film *Syriana* which is about the oil industry. It highlights political and corporate corruption around the industry, but the central theme is America's dependence on foreign oil. Most recently, the movie *The Kingdom* was released as a feature film about terrorism and anti-American sentiment in the Middle East which was closely linked to oil. More and more, the issue of petroleum security is taking root in the nation's psyche, and the threat of a shortage is becoming an increasingly widespread concern, and rightfully so.

By 2010, an increase of approximately 46 percent above existing capacity production rates will be necessary to meet projected demand. Production from existing reserves will decline while demand will increase, and by 2020, the requirement to meet demand will be the significant addition of new production. **Unfortunately, the further development of existing reserves, combined with enhanced oil recovery techniques, will fall short of meeting the future demand for oil.**

As demand for energy increases, oil and gas should continue to dominate the energy mix. Unfortunately, the addition of new oil reserves has decreased over the last 40 years while the world continues to rely upon mature oil fields to meet the growing demand. Financial markets appear to support an overall commodity based shift of which oil and gas are major components. New oil and gas discoveries will be necessary to meet future demand requirements, but it is questionable where these discoveries will be made. As discussed earlier, developing nations such as India and China are actively seeking energy to realize their full economic growth potential. This has made global E&P activities highly competitive and politically charged. Given the current geopolitical climate and global attitudes towards the United States, it is in the interest of U.S. national security to further develop their domestic reserves. Succinctly:

- The world's demand for energy is rapidly increasing, and

- geopolitical tensions in hydrocarbon rich areas of the world make the United States economically vulnerable.

In fact, the International Society of Energy Advocates (www.energyadvocates.org), an educational and advocacy group for the U.S. energy industry, is currently running a national education campaign with the slogan **America Needs America's Energy.**

Over 200 billion barrels of unrecovered oil lies beneath the U. S. at shallow depths of 5000 feet or less; however, the disbursement of these reserves is wide, and reserves-per-well estimates make the production of these shallow wells marginally profitable. However, these complicated drilling programs, when combined with the high costs, make these upstream opportunities appear less attractive to energy investors. Unfortunately, technology had not been available to make shallow-well or tight prospects economically feasible. Other challenges, similar to those of shallow oil, plague the development of alternative fuels sources as well. Often the dominant downfall of alternative energy is economics. Either the production of energy is too costly to justify, or the energy required to produce the alternative form of energy is greater than the energy produced.

XIV
The State of Alternatives

In the United States, transportation is projected to be the major use for oil by 2015. This forecast by the EIA and the petroleum security issues that have threatened America for the last 30 years have spurred the discussion about the development of alternative fuels.

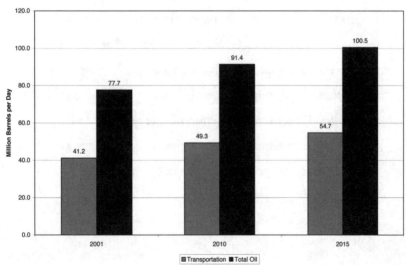

Figure 10—Source: EIA

The State of Alternatives

In the United States, renewable/alternative energies make up approximately 6 percent of the total energy consumed. It should be noted that biomass and hydroelectric account for nearly all of the alternative fuels. Solar, wind, and geothermal alternatives contribute less than 10 percent to the energy mix.

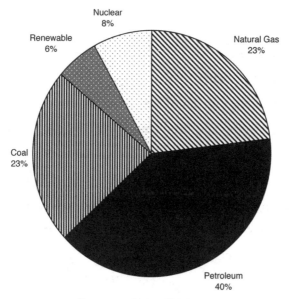

Figure 11—Source: EIA Data

The significance of biomass in alternative energy is largely attributable to ethanol. There was much talk in the early 2000s about expanding the use of ethanol as an alternative to gasoline. When compared with conventional unleaded gasoline, ethanol is a particulate-free fuel source that combusts with oxygen to form carbon dioxide and water. Today, ethanol is routinely used as a blend stock for gasoline. As of 2006, the feedstock for producing ethanol comes primarily from sugarcane, maize (corn) and sugar beets. Four countries have developed significant bioethanol fuel programs:

- Brazil
- Colombia
- China
- United States

Of these four countries, Brazil is the only country where farming and production of ethanol is a profitable and widespread substitute for gasoline. Nevertheless, there are still inefficiencies in using ethanol, as well as other alternative sources as fuel. The following matrix describes the relative advantages and disadvantages of the most prominent alternative fuels.

The State of Alternatives

	Octane Number	Main Fuel Source	Energy Content per Gallon	Energy Ratio Compared to Gasoline	Types of Vehicles Available today
Gasoline	86 to 94	Crude Oil	109,000 – 125,000 Btu		All types of vehicle classes
Biodiesel (B20)	-25	Soy bean oil, waste cooking oil, animal fats, and rapeseed oil	117,000 - 120,000 Btu (compared to diesel #2)	1.1 to 1 or 90% (relative to diesel)	Any vehicle that runs on diesel today—no modifications are needed for up to 5% blends. Many engines also compatible with up to 20% blends
Compressed Natural Gas (CNG)	120+	Underground reserves	33,000 - 38,000 Btu @ 3000 psi; 38,000 - 44,000 @ 3600 psi	3.94 to 1 or 25% at 3000 psi; 3.0 to 1 @ 3600 psi	Many types of vehicle classes.
Electricity	N/A	Coal; however, nuclear, natural gas, hydroelectric, and renewable resources can also be used.	N/A		Neighborhood electric vehicles, bicycles, light-duty vehicles, medium and heavyduty trucks and buses.
Ethanol (E85)	100	Corn, grains, or agricultural waste	~80,000 Btu	1.42 to 1 or 70%	Light-duty vehicles, medium and heavyduty trucks and buses — these vehicles are flexible fuel vehicles that can be fueled with E85, gasoline, or any combination of the two fuels.
Hydrogen	130+	Natural Gas, Methanol, and other energy sources.	Gas: ~6,500 Btu@3,000 psi; ~16,000 Btu@10,000 psi Liquid: ~30,500 Btu		No vehicles are available for commercial sale yet, but some vehicles are being leased for demonstration purposes.

Figure 12a —Source: EIA

The Braking Point

	Environmental Impacts of Burning Fuel	Energy Security Impacts	Fuel Availability	Safety Issues (Without exception, all alternative fuel vehicles must meet today's OEM Safety Standards.)
Gasoline	Produces harmful emissions; however, gasoline and gasoline vehicles are rapidly improving and emissions are being reduced.	Manufactured using imported oil, which is not an energy secure option.	Available at all fueling stations.	Gasoline is a relatively safe fuel since people have learned to use it safely. Gasoline is not biodegradable though, so a spill could pollute soil and water.
Biodiesel (B20)	Reduces particulate matter and global warming gas emissions compared to conventional diesel; however, NOx emissions may be increased.	Biodiesel is domestically produced and has a fossil energy ratio of 3.3 to 1, which means that its fossil energy inputs are similar to those of petroleum.	Available in bulk from an increasing number of suppliers. There are 22 states that have some biodiesel stations available to the public.	Less toxic and more biodegradable than conventional fuel, can be transported, delivered, and stored using the same equipment as for diesel fuel.
Compressed Natural Gas (CNG)	120+ CNG vehicles can demonstrate a reduction in ozoneforming emissions compared to some conventional fuels; however, HC emissions may be increased.	CNG is domestically produced. The United States has vast natural gas reserves.	More than 1,100 CNG stations can be found across the country. California has the highest concentration of CNG stations. Home fueling will be available in 2003.	Pressurized tanks have been designed to withstand severe impact, high external temperatures, and automotive environmental exposure.
Electricity	Efficient Vehicles have zero tailpipe emissions; however, some amount of emissions can be contributed to power generation.	Electricity is generated mainly through coal fired power plants. Coal is the United States' most plentiful fossil energy resource and our most economical and pricestable fossil fuel.	Most homes, government facilities, fleet garages, and businesses have adequate electrical capacity for charging, but, special hookup or upgrades may be required. More than 600 electric charging stations are available in California and Arizona.	OEM EVs meet all the same vehicle safety standards as conventional vehicles.
Ethanol (E85)	E85 vehicles can demonstrate a 25% reduction in ozoneforming emissions compared to reformulated gasoline.	Ethanol is produced domestically and it is renewable.	Most of the E85 fueling stations are located in the Midwest, but in all, approximately 150 stations are available in 23 states.	Ethanol can form an explosive vapor in fuel tanks. However, ethanol is less dangerous than gasoline because its low evaporation speed keeps alcohol concentration in the air low and non-explosive.
Hydrogen	Zero regulated emissions for fuel cell-powered vehicles, and only NOx emissions possible for internal combustion engines operating on hydrogen.	Hydrogen can help reduce U.S. dependence on foreign oil by being produced by renewable resources.	There are only a small number of hydrogen stations across the country. Most are available for private use only.	Hydrogen has an excellent industrial safety record; codes and standards for consumer vehicle use are under development.

Figure 12b —Source: EIA

At no other time in world's history has the quest for energy been more diligent. As we have seen, the tight supply of oil the world is currently experiencing has already placed upward pressure on oil prices. However, higher prices make otherwise economically unviable alternatives a plausible option, as we have seen in the case of Oil Sands.

Oil Sands

Until recently, Alberta's bitumen deposits were known as tar sands but are now called oil sands. Oil sands are deposits of bitumen—viscous oil that must be rigorously treated in order to convert it into an upgraded crude oil before it can be used in refineries to produce gasoline and other fuels. Bitumen is about 10-12 percent of the actual oil sands found in Alberta. The remaining 80-85 percent is mineral matter, including clay and sands, and around 4-6 percent water. While conventional crude oil is either pumped from the ground or flows naturally, oil sands must be mined or recovered *in situ* (meaning *in place*). The oil sands recovery process includes extraction and separation systems to remove the bitumen from the sand and water.

Oil sands currently represent 40 percent of Alberta's total oil production and about one-third of all the oil produced by Canada. Soon, oil sands production is expected to represent 50 percent of Canada's total crude oil output and 10 percent of North American production. Although tar sands occur in more than 70 countries, the two largest are Canada and Venezuela, with the bulk being found in four different regions of Alberta, Canada: areas of Athabasca, Wabasha, Cold Lake and Peace River. In fact, the reserve that is deemed to be technologically retrievable today is estimated at 280-300 GB (billion barrels). This is larger than the Saudi Arabian oil reserves, which are estimated at 240 GB. The total reserves for Alberta, including oil not recoverable using current technology, are estimated at 1,700- 2,500 GB. There are 174 billion barrels of established reserves. Alberta's oil sands rank as the world's largest source of crude oil outside Saudi Arabia.

With the United States' decline in domestic oil production, it is important to research possible oil sand production in Alaska. Our consumption continues to increase along with our dependence on oil imports. The deposits of oil sands (oil shale) in the United States are massive. The processing of oil shale has gone through cycles of development and commercialization, all without achieving a competitive cost of production. As well, tar sands are processed on a limited basis. An engineering study was done by the University of Alabama and the Department of Energy that provides a preliminary design of a commercial processing facility to beneficiate 39,956 tons per day of run-of-mine eastern oil shale to produce 4.38 million tons per year of concentrate. The report includes a process for recovery of kerogen at 92 percent, which with hydroretorting would produce approximately 20,000 barrels of oil a day.

Though these estimates are promising, oil sands alone will not remedy the world's thirst for energy. In fact, fuel for transportation is only part of the equation—the other challenge the world faces is power generation.

Power Generation

Following a presentation in Cushing, Oklahoma (the pipeline crossroads of America) on the energy challenges facing America, a person in the audience stated to me, "I don't know what you are worried about. If we run out of oil and gas we'll still have electricity." Unfortunately, this is a widespread misconception. With the exception of lightening, and some very small amount of static, there are no naturally occurring forms of electricity. Power generation requires the use of other energies and just as renewable/alternative energies comprised a minority share of fuels for transportation, the same is true of power generation.

Power Generation by Source

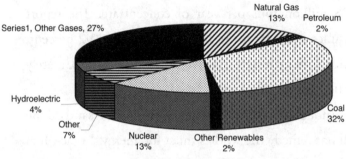

Figure 13—Source: EIA Data

Nonetheless, various means of power generation are touted as the next *cure*, and the most popular are wind and nuclear fission. Many have asked why we do not simply rely on wind power as a main source of energy. While wind generation of electricity is clean and renewable, it is highly unreliable. Wind patterns and frequency cannot be easily predicted. Additionally, the costs associated with wind generation prohibit it's widespread application. Power generation through wind is economic only in the very highest price environment. Unfortunately, this is the case with many *alternative* energies.

However, unlike other forms of power generation, nuclear power plants use the heat produced by fission to generate electricity. Approximately, 20 percent of the total net electricity generated in the United States is done so with nuclear power. This is close to the amount of electricity used in California, Texas and New York, the three most populous states. Yet there are other, non-economic concerns about the expanded use of nuclear power generation, and they include the following:

- Radioactive Waste
- Possible targets of terrorism
- Finite supply of Uranium

Both the existing and potential side-effects of expanded nuclear power generation seem almost counter productive with regard to the environment, stabile uranium supply, and national security.

Despite the overwhelming evidence that *alternatives* will not free the world from our *hydrocarbon chains*, there is a promising idea for power generation that utilizes our abundant supply of coal. This concept, supported by the U.S. Department of Energy, is FutureGen.

FutureGen

Coal is the least costly and most abundant energy resource available in the U.S., so FutureGen, a $1 billion government-industry project, was proposed to make the most of this plentiful fossil fuel. Essentially, the goal is to provide a first-of-its-kind clean power plant, with a target completion date of 2012. Nearly every facet of the prototype plant will be based on cutting edge technology that already exists. In fact, the IEA projects coal to be the dominant fuel type for power generation through 2030.

The highly energy efficient coal supplied power plant should produce near zero emissions. The power plant is to be designed to generate hydrogen and electricity using a process called carbon capture and storage (CCS). This is an approach to mitigating climate change by capturing carbon dioxide (CO_2) from large point sources such as power plants and subsequently storing it away safely instead of releasing it into the atmosphere. This will reduce common air pollutants like sulfur dioxide and nitrogen oxides and convert them to useable by-products such as fertilizers.

So far, no one method has proven to be a measurable addition to the energy mix. Despite our technological advances, the economics of producing alternative fuels are challenging. In fact, the IEA projects alternative fuels to account for less than 10 percent of the total energy mix by 2030. Therefore, conventional energies are expected to remain the dominant sources of fuel for some time.

XV
The Second Phase of the Hydrocarbon Age

As the world struggles to produce the fuel to power the advancement of humankind, there will be tensions between nations with energy reserves and those without. War is a costly manner in which to procure necessary resources, but peace can eradicate supply problems. **However, can there be peace? Furthermore, despite peace, will there be enough resources to satisfy the world's thirst for oil?**

In both of these cases, the answer may be no. The world attempting to procure resources at a frenzied pace, and this competition is an obstacle to peace. As for oil, it seems we are approaching the sunset of its tenure as the dominant fuel. **However, this sunset may lead to the rise of another fuel—natural gas.**

The future energy will not be the black gold America has come to romanticize over the last 100 years—rather natural gas looks to be the bridge between oil and the unknown energy of the next century.

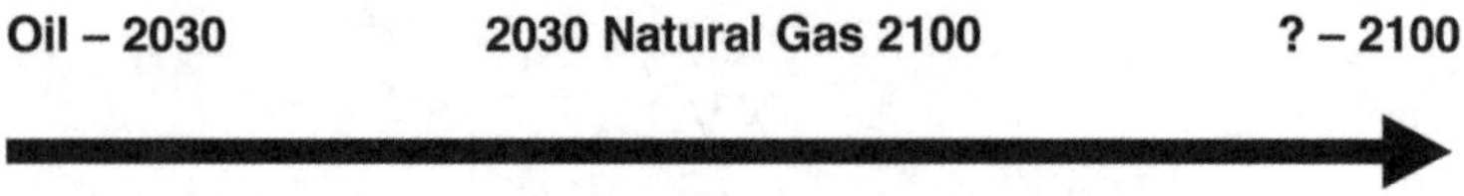

```
Oil – 2030        2030 Natural Gas 2100        ? – 2100
━━━━━━━━━━━━━━━━━━━━━━━━━━━━━━━━━━━━━━━━▶
```

Figure 14

First, we have to get to 2030, and oil is clearly the vehicle to get us there. Peace will be an important factor. Given the events of late, it is difficult to imagine peace in the Middle East. Nevertheless, can Iraqi oil bring peace to the unstable Middle East? If so, how will this affect Iraq's neighbor and the world's largest producer of oil—the Kingdom of Saudi Arabia?

At the moment, eyes on Iraq are focusing on terrorist insurgency, and recently, it has been cited as the reason for limited Iraqi oil production. However, once the nation is able to fully develop their immense reserves, this brutal terrorism should diminish in the wake of economic prosperity. Bringing Iraqi crude to market, especially in a climate of rising prices, will serve the reconstruction of the nation and its economy well. Furthermore, the continued development of Iraqi oil is not only in the best interest of Iraq, but also of Saudi Arabia.

Saudi Arabia has much to gain by peace in the Middle East, and their continued prosperity in the global oil market may be linked to the development of Iraqi oil.

Additional reserves brought to market by an increase in Iraqi production could appear as a threat to the recent upward pressure on the price of crude, but it is estimated that the growing demand for oil from developing nations, such as China and India, will easily absorb additional supply. Over time, the continued development of these Asian economies will support the trend of increasing demand; therefore, crude prices should remain relatively high, and significant economic development will be enjoyed by the oil producing nations of the Middle East—thus the benefit of producing Iraqi oil reaches beyond its borders.

OPEC's oil embargo during the 1970s resulted in the North Sea discovery, and by 1982, non-OPEC production surpassed OPEC production by 1 million barrels per day. Today, the efficient production of unconventional oil and the evolution of alternative/renewable fuels will flourish in the event of high prices influenced by the artificial manipulation of supply; if this takes place, the ultimate economic loser will be the Middle East. Therefore, the tightening of supply, whether by OPEC manipulation or real demand, is not in the best interest of the oil producing nations of the Middle East, and nothing is gained by impeding the development of Iraqi reserves.

Iraq can be a key participant in the global community,

and unfortunately, terrorism is obstructing their path to future economic development. The circumstance of increasing demand for oil, the competition of unconventional oil production, and the development of alternative/renewable fuels will require a peaceful resolution to differences in the Middle East if prosperity is to be realized. For the producing nations of the world, peace will be a necessity if they are to be leaders in the second phase of the Hydrocarbon Age.

Recently, we had the opportunity to speak with leading oil analyst Mr. Charley Maxwell. Charley has been a voice in the industry for over 30 years, and he was ranked by Institutional Investor as the number 1 oil analyst for much of the 70s and 80s. Today, much of Mr. Maxwell's work is focused on the coming oil crisis. He meets semiannually with OPEC to discuss near-term and long-term planning. We were interested in learning Mr. Maxwell's insights to the energy conditions today and his expectations of the next 30 years.

Q1. *Geopolitical tensions are high in regards to the global quest for oil. What can we expect in the coming years as demand begins to outstrip supply?*

A1. Well, in current economic terms, I think that we

will find that while an individual nation can set in price controls, as we attempted to do in this country under Carter's administration, most nations competing in the world market cannot, and it will be a *free for all* epic market for oil. Therefore, the mechanism that will work to ration oil to supply is rationing by price.

Other than some small variations in inventory, supply must meet demand. If we have less supply and demand continues to rise, and both are likely to happen, then we could have continuous price increases. For example, if demand would normally rise by 1.5 percent over the next 20 years, but production declines by 0.5 percent then we have a supply/demand gap of 2 percent, and this 2 percent must be replaced. I think the elasticity of oil is such that it will require immense price increases to price out 2 percent of world demand.

Q2. *We have already seen some weakening of the OPEC "oil muscle." How much of a punch do they have left?*

A2. It varies at times. For instance they couldn't limit the upward price of oil during the summer of 2006 because they were at full production. OPEC's ability to limit prices increases has shown to be very modest. You don't often think of OPEC trying to limit the price of oil, but they did because they felt too high of a price would be dangerous—and it was. They would prefer $60 oil, but it ended up being $78. This demonstrates they do not have much upside control. However, we are not often worried about upside control; rather, we are more concerned with OPEC's ability to influence prices on the downside.

The present conditions affecting OPEC are as follows:

    1. OPEC is trying to preserve their revenue base, which is oil. Right now, it is more advantageous to cut production and allow prices to rise than it is to increase production and sell more oil at a lower price. These are difficult times for OPEC, and they're at risk of losing price control on the downside.

    2. OPEC is under some pressure to stay together and act as a group as opposed to individual nations. In the past, when any given OPEC member had financial problems, they would cheat on the other member countries, but they have had some good prices in the last couple of years, and their treasuries are well supplied. They now have the money to make it through the cuts in production. OPEC is much more

sophisticated now, and they understand acting together is in their best interest.

3. Saudi Arabia is very important to OPEC and, having the greatest production capacity, they are the only swing producer in the organization. They have a very intelligent and personable leader now, Ali I. Al-Naimi, the Saudi Oil Minister; OPEC looks to Al-Naimi for leadership, and he is perfect for this job. He is knowledgeable about this business and understands the power of Saudi Arabia and the limits of their power. He argues points quietly, rationally, and effectively. He has a very important role in helping OPEC avoid disconformity amongst members.

Q3. *Speaking of disconformities, occasionally there is talk of trading oil in euros rather than dollars. If this were widely implemented, what would the impact be on the U.S. and the rest of the world?*

A3. Well, the lower the dollar falls the higher the price. Recently Art Laffer has made statements that he believes the price of oil will go down to $30 over the next two years; I don't agree with him. His projection was based on the idea that high prices would spur demand destruction while providing an incentive for oil companies to increase production. However, the reason we are not producing more oil has little to do with the price. The reason we are producing less oil is because we don't have access to places where oil is still available to be produced. Yet, if a lot of people believe the price of oil is going down it can have a certain momentum. The market will begin to sell oil forward because they see it as an opportunity to sell short. I usually talk about supply and demand as being the main determinant in price, and it is, however the effect of large amounts of money trading around the world can cause a psychological shortage or surplus of oil.

Q4. *Will production of non-conventional oil fill the gap before widespread application of other alternatives?*

A4. Each alternative fuel source makes its own contribution. However, I think these will contribute a very slight amount of energy due to the slim margins they yield. You have to be careful when considering alternative fuels because of the energy it takes to produce them; this is the curse of alternative energy. Not to mention that, on a macro scale, alternative energy production can adversely affect manufacturing capacity because we are allocating capacity to energy production instead of making more widgets.

Q5. *With natural gas relatively abundant, do you see a time when NG could rival oil as the dominant energy supply?*

A5. Well, yes of course. It is estimated that world gas production won't peak until somewhere between 2030 and 2045 while oil is projected to peak between 2015 and 2020. **So, oil should remain dominant until around 2030 when natural gas could surpass oil.**

Q6. *For this to happen, Liquefied Natural Gas (LNG) distribution will be critical. Where do you see investment dollars coming from?*

A6. LNG investment dollars will be funded by the majors; it's their kind of deal. It involves large project management, immense capital, and access to large gas reserves abroad and relationships with the nations that house those reserves. All of these strengths allow the majors to bargain for the gas at a price that can make LNG economic. I believe the majors decided along time ago LNG was the way to go. You can see this by the way they shifted their focus away from smaller American reserves to larger international discoveries. They sold off their assets domestically and mid-sized oil companies thrived.

Q7. *What will the role of independents be in the future* *and how will things change for them?*

A7. There will still be some large oil and gas discoveries in the U.S.A. like the Wolverine discovery in Utah. So, there will be opportunity for the independents to come in and continue developing our domestic reserves, but I don't believe many will want to participate, and the reason is due to the price that will occur due to LNG. The geology that is left in the United States just isn't promising enough to warrant significant exploration activity. The geology has been picked over and over.

Q8. *Bottom-Line*

A8. Oil prices will fluctuate with the availability of oil, and these prices will help to knock out some demand, but not much. Everyone in the third world wants to ride—not walk. Therefore, the demand side of the equation does not have many possibilities of letting up, and the supply side has significant impediments as well.

> 1. 75-80 percent of the world's oil is produced by National Oil Companies (NOCs) and we cannot force their hand. In many cases, these companies are inefficient, they cannot attract top talent, and there is little incentive for them to increase E&P activity which would contribute to

the lowering of prices.

2. The political instability of many of the nations which still possess significant reserves is a primary deterrent for private oil companies to go in and develop these resources.

3. The majors have exhibited a lack of vision for their industry. You have seen companies use their profits to buy back stock rather than exploring and adding to their reserves. The easy answer for them is, "there is nowhere to drill," but I am not convinced this is entirely accurate.

4. Geologic alternatives are shrinking. This will be more of an impediment in the future, but eventually, we will have accessed all of the most significant discoveries.

The result: Supply and demand will be in rough equivalence in 2007.
(End of Interview)

In rough equivalence indeed. Both China and India show no signs of slowing in economic growth, and the United States is still on course to being the largest consumer of energy in the foreseeable future—**enter natural gas.**

In the United States, natural gas is still relatively

abundant, but the problem with gas has always been mobility. Unlike oil, which is easily transported, it is much more difficult to move gas from the source to the consumer. The additional cost of pipelines and processing hinders the economics of producing gas. However, the relative high price of oil and the increasing price of gas should help to break the economic barriers restricting the continued development of America's stranded reserves. This has already begun to take place as the nation has begun accelerating the development of non-conventional oil plays such as coal bed methane (CBM) and shales.

The concept of a natural gas fueled country makes sense on several levels. First, natural gas is more abundant than oil, and it is abundant domestically. Automatically, the U.S. becomes less dependent on the unfriendly foreign sources of energy. Next, natural gas is far easier on the environment than oil. Lastly, progress takes the path of least resistance, and a shift to natural gas will most likely be met with the fewest policy changes and industrial resistance. However, the transition to natural gas as the dominant source of energy will not come without some effort. Reinvestment in our infrastructure and our oil and gas industry will be critical for future success.

Iran has the second largest estimated reserves of natural gas in the world. In late 2006, we had the opportunity to discuss the important role Iran will play in helping the world meet its future energy demand with the former crown prince of Iran, HRH Reza Pahlavi.

Q1. *Despite strong revenues from oil exports, Iran has been unable to modernize its oil and gas sector. What are the key factors inhibiting the further development of Iran's most vital industry?*

A1. Gradually, (after the regime change) the political process continued down to the lower levels of the organization and into operational units. At the same time the government unilaterally annulled all existing exploration and production agreements with international oil companies cutting the inflow of foreign capital as well as modern technology which the industry severely needed.

Lack of adequate investment and application of sound oil industry practices and advanced technologies which was the consequence of loss of experienced professionals as well as economic and political isolation coupled with chronic mismanagement inhibited further modernization and development of Iran's oil and gas industry.

Q2. *What changes in U.S. policies could be made to influence regime change in Iran and reestablish foreign investment in the Iranian energy sector?*

A2. The so-called Islamic Revolution was about 28 years ago, and two entire generations have grown up since. The result is that about 70 percent of Iran's population is under 30, and they are deeply dissatisfied with the unequal distribution of oil wealth and the dysfunctional economy. In large part, they long for the inalienable rights and the material privileges that the West takes for granted. While there is intense national pride, there is very little evidence of the ideological fervor that one would think is rampant in Iran. Further, there is a massive managerial vacuum in Iran due to the hierarchical and bureaucratic decision-making structure and the fact that many decision-makers are not necessarily appointed for their managerial competence. Consequently, discrimination and favoritism in society certainly could provoke retaliation among the youth.

Reestablishment of foreign investment in the Iranian energy sector can only be achieved with the creation of: Firstly, an environment of political and economic stability and transparency in the country and secondly, an efficient, professional, qualified and honest management in the oil and gas industry.

Q3. *Natural gas will become increasingly important to meet the future global demand for energy, and Iran has an estimated 970 Tcf of proven gas reserves. Expanded LNG operations could be very favorable for the Iranian economy, but the cost of developing a network of terminals will be immense. Where can LNG investment dollars be acquired in both the near and long term?*

A3. As is now common knowledge among petroleum industry players, gas is gaining significant shares in the global energy mix, and the World Bank estimates that world demand for gas will exceed that of oil by 2025. Luckily, some 974 Tcf (*Oil & Gas Journal,* though this might be an inflated figure) of world gas reserves are also domiciled in Iran, second only after Russia.

Whereas supply from long-distance sources seemed unthinkable two decades ago, it has been made possible by new technologies that allow the conversion of natural gas to liquids for shipment in tankers as well as improvements in pipeline technology. Now, much of the spur for gas demand comes from the fact that it is the most environmentally friendly of fossil fuels. Therefore, an increasing number of countries are turning to gas, mostly for power generation but also for traditional domestic use.

The two most advanced technologies for conversion of natural gas to liquids for long haul shipment to major consuming markets are LNG (Liquefied Natural Gas) and GTL (Gas to Liquid Products). Unfortunately due to current adverse economic and political situations Iran with its vast reserves of gas is facing tough competition from Qatar for Far East and Asian markets and from Russia, Kazakhstan and Algeria for European markets. Unlike Iran, all these countries have succeeded in attracting the capital, technology and cooperation of international oil companies in developing LNG and or GTL projects for export.

Q4. *As natural gas becomes a significant export for the Middle East, Iran's immense natural gas reserves could position the country to take a leading role in both OPEC and the region. In what ways could Iran be the catalyst for revolutionizing the energy trade in the Middle East?*

A4. As oil is an important strategic global commodity and affects daily life everywhere, a secular and democratic Iran must in the future take the lead to firstly prevent OPEC from ever using oil or gas as a political weapon and secondly to encourage OPEC member countries in organizing an effective and useful dialogue with oil consuming countries.

Iran should use its influence to encourage OPEC to promote the spirit of *partnership* with consuming countries and international oil companies so that the *energy supply security* so vitally needed by consumers could go hand-in-hand with *energy demand security* essential for producing countries to invest safely and develop their production capacities to meet the world oil and gas demand. (END)

As we move towards a natural gas powered world, it seems clear that countries like Iran will play a significant role in helping the world meet its energy needs in the future.

XVI
The Courage to Lead

In proven reserves, the world has an estimated 40 years of oil and 60 of natural gas left. However, many of America's remaining reserves are located in off limits areas.

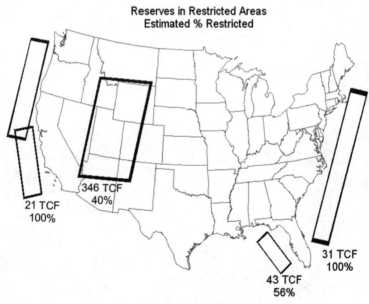

Figure 15—Source: EIA

These are America's resources, and with responsible production, these resources can play an important role. The energy question has never been more critical than it is today. As we have witnessed since the 1970s the petroleum security of the United States, and the world, is at the forefront of political policy.

The United States must put forth effort to address our oil consumption, but the solution is not to tax our way out of the crisis. In late 2006, California attempted to pass legislation that would finance research for alternative energy with tax revenues generated from oil and gas production in the state. The following outlines Proposition 87.

ALTERNATIVE ENERGY. RESEARCH, PRODUCTION, INCENTIVES. TAX ON CALIFORNIA OIL PRODUCERS. INITIATIVE CONSTITUTIONAL AMENDMENT AND STATUTE.

- establishes $4 billion program with goal to reduce petroleum consumption by 25%, with research and production incentives for alternative energy, alternative energy vehicles, energy efficient technologies, and for education and training
- funded by tax of 1.5% to 6% (depending on oil

price per barrel) on producers of oil extracted in California. Prohibits producers from passing tax to consumers
- program administered by new California Energy Alternatives Program Authority
- prohibits changing tax while indebtedness remains
- revenue excluded from appropriation limits and minimum education funding (Proposition 98) calculations

Summary of Legislative Analyst's Estimate of Net State and Local Government Fiscal Impact:
- new state revenues—depending on the interpretation of the measure—from about $225 million to $485 million annually from the imposition of a severance tax on oil production, to be used to fund $4 billion in new alternative energy programs over time
- potential reductions of state revenues from oil production on state lands of up to $15 million annually; reductions of state corporate taxes paid by oil producers of up to $10 million annually; local property tax reductions of a few million dollars annually; and potential reductions in fuel-related excise and sales taxes

Source: California General Election Official Voter Information Guide

The proposition failed to pass, and it is clear that taxation is not the answer. This nation's last three recessions have all been linked to rising energy prices — and there is strong evidence that the latest round of escalating energy prices is having a negative effect on the longest period of prosperity in the country's history. The National Association of Manufacturers estimates that the soaring fuel prices between 1999 and 2000 cost the U.S. economy more than $115 billion. During a two week period last January, Californians lost an estimated $2.3 billion in wages, sales and productivity. Layoffs are already hitting workers in the Western United States, as companies shift production to states with more reliable energy sources.

The power crisis isn't just hitting our pocketbooks — it's changing the way we live our lives.

What do we need to do to reorganize this problem?

- Adopt a comprehensive national energy policy.
- Lift the regulatory impediments to the development of additional sources of oil and gas.
- Encourage the development of new generating facilities in areas of short supply.
- For an example, standardize the emission control laws so gasoline refined in Oklahoma can be sold in Illinois.

- Encourage the investment in new refineries, transmission line and pipelines through governmental cooperation with companies willing to invest the billions necessary for their construction.
- Allow the free market mechanism to bring supply and demand into equilibrium at a competitive price.

In October 1999, I was invited to *The San Antonio Round Table* sponsored by The Institute for the Study of Earth and Man, SMU. The roundtable included twenty individuals from across the globe. We discussed views of the world ahead:

- the lasting impact of crude price collapse and subsequent recovery
- continuing changes in industry structure
- asset management
- political change and opening in the Middle East
- an independent's strategies
- managing human resources in a rollercoaster industry
- the Caspian-continued promise or disappointment
- revolution or evolution in transportation fuels

- nuclear fusion
- gas to liquids
- global warming

Among those at the table were, Nick De'Ath of Strategic Risk Services, Price Waterhouse Coopers, Leo Drollas, Deputy Executive Director Chief Economist of the Centre for Global Energy Study, Anthony Finizza, Economic Counselor for ARCO, Peter Gaffney, Senior Partner of Gaffney, Cline and Associates, Herbert Hunt of PetroHunt Corporation, Thomas Meuer, President of Hunt Oil Co., Edward Morse of Hess Energy Trading, and Frank Sprow, VP of Exxon Corporation. Now, much like October 1999, the discussion of a global energy crisis is underway, and this is a promising step in the right direction. We need the courage to lead!

Activism

The International Society of Energy Advocates organization, www.energyadvocates.org, was founded in 1974, and it was established due to the 1973-74 OPEC oil embargo. It was during this time in history that a group of oil executives formed the organization as a not-for-profit national energy education organization. The Energy Advocates' primary mission is to inform the general public about our vital energy industry and energy policy issues.

Since 1974, members of The Energy Advocates have spoken in all fifty states, appeared on television, spoken on radio, and have been quoted in newspapers and magazines on behalf of the energy industry.

The organization believes that it is important that we rise to the challenge and make a difference when it comes to energy issues and energy public awareness. One way to bring attention to these issues was through the formation of the International Energy Policy Conference. The first conference was held at the University of Oklahoma in 1992, and since then, there have been fifteen conferences held in the United States including Washington, D. C., Denver, Dallas, Oklahoma City, and Tulsa.

The Road Ahead

In the 1980s, the United States passed laws that prohibited the construction of any more electrical generating plants that used natural gas as primary fuel. We did not allow any new industrial plant to rely on natural gas as its primary source of heating etc. and we began spending billions of dollars trying to develop alternative fuels and sources such as shale oil. At the time, natural gas prices had been under strict price controls for many years and were between $.20 and $.40 Mcf. Because of the risk of drilling below 15,000 feet, Congress allowed any gas found below this level to be sold at whatever price it could command. Exploration companies and others spent billions of dollars looking for and finding deep gas. The result? Trillions of cubic feet of gas deposits were identified and brought on line. The result after some wide swings in prices in the late 80s was a huge surplus of gas for over ten years.

Natural gas will be very significant in the years ahead however, oil is still our major energy source, and it also presents us with our greatest challenge. **It will take courage to make the critical and difficult decisions that challenge the way we consume and produce energy**. In our very recent history, America and the world has witnessed explosive progress in both science and technology—the knowing and the doing. However, this happened so fast that our development outstripped resources quicker than we would have ever imagined. The greatest advances of humankind have occurred in the last 150 years, and those advances were fueled by oil. Now, we must discover a way to continue advancement without it. We realize we are globally interdependent on energy supplies. Therefore, we must continue to develop relationships with our foreign allies.

For more information on what you can do, go to:

www.thebrakingpoint.com

There you will find:

- updates on energy policy
- energy saving tips
- resources for industry, government, and consumers
- self energy audit

Be part of the solution. We are at *The Braking Point*. Let the debate begin!

Appendix I

THE WHITE HOUSE
Office of the Press Secretary
For Immediate Release
July 29, 2005

The Energy Bill: Good For Consumers, The Economy, And The Environment

"America must have an energy policy that plans for the future, but meets the needs of today. I believe we can develop our natural resources and protect our environment."

– President George W. Bush

FACT SHEET

President Bush entered office calling on Congress to pass the first national energy plan in a generation. He proposed a comprehensive energy plan to encourage conservation and energy efficiency; expand the use of alternative and renewable energy; increase the domestic production of conventional fuels; and invest in modernization of our energy infrastructure.

The energy bill passed by Congress this week paves the way for a brighter and more secure energy future with more reliable, affordable, and clean sources of energy to power America forward. It will help put us on the path to reducing our dependence on foreign sources of energy. Our reliance on imported energy did not come about overnight, and it will take time to reverse.

By harnessing the power of American innovation and technological development, the energy bill will help us transform the way that we use and produce energy — resulting in greater energy security, a growing economy, and a healthier environment for generations of Americans to come.

To Encourage Conservation And Energy Efficiency, The Energy Bill:

- Establishes new energy efficiency standards for a wide variety of consumer products and commercial appliances, and offers tax incentives to encourage their purchase.

- Encourages improved efficiency in homes and buildings, establishes new aggressive Federal energy savings goals, and reauthorizes the Energy Savings Performance Contract program to conserve more energy at Federal facilities.

- Offers tax incentives to consumers to purchase energy-efficient hybrid, clean diesel, and fuel cell vehicles.
- Requires a new, multi-year rulemaking by the Department of Transportation to increase fuel economy standards for passenger cars, light trucks, and SUVs.

To Expand The Use Of Alternative And Renewable Energy, The Energy Bill:

- Establishes a new Renewable Fuel Standard that requires the annual use of 7.5 billion gallons of ethanol and biodiesel in the nation's fuel supply by 2012
- Extends the existing tax credit for production of electricity from renewable resources, such as wind, biomass, and landfill gas, and creates for the first time a tax credit for residential solar energy systems
- Authorizes full funding for the President's Hydrogen Fuel Initiative
- Provides Federal risk insurance and extends

the Price-Anderson Act to mitigate the potential cost of unforeseen delays and encourage investment in a new generation of safer, more reliable, and more proliferation-resistant nuclear power plants.

To Increase The Domestic Production Of Conventional Fuels, The Energy Bill:

- Makes needed reforms to clarify the onshore oil and gas permitting process, and reduce conflicts with other laws and regulations (stormwater, CZMA, hydraulic fracturing)
- Clarifies FERC jurisdiction over siting of onshore LNG facilities to accelerate development of a global market in natural gas and help reduce prices for U.S. consumers
- Authorizes full funding for the President's Clean Coal Research Initiative and updates Federal coal leasing laws
- Eliminates the 2 percent "oxygenate requirement" for reformulated gasoline, to improve the flexibility of our fuel supply and reduce the number of *boutique fuels*.

To Encourage Investment In Modernization And Reliability Of Our Energy Infrastructure, The Energy Bill:

- Requires mandatory reliability standards to make the electric power grid more reliable and protect against blackouts
- Reforms outdated tax laws to expand investments in electric transmission and generation facilities
- Establishes last-resort Federal siting authority for transmission lines deemed in the national interest to ensure a better functioning power grid.

The Energy Bill Also Helps Reduce The Global Demand For Energy By:

- Working with our international partners—including fast growing nations like China and India—to encourage them to deploy the cleanest and most efficient energy technologies as they develop and grow their economies.

The Braking Point

Appendix II

Appendix II

Reprinted from the EIA website. "This chronology was originally published by the Department of Energy's Office of the Strategic Petroleum Reserve, Analysis Division. Updates for 1995-2005 are from the Energy Information Administration." (EIA)

World Nominal Oil Price Chronology: 1970-2005

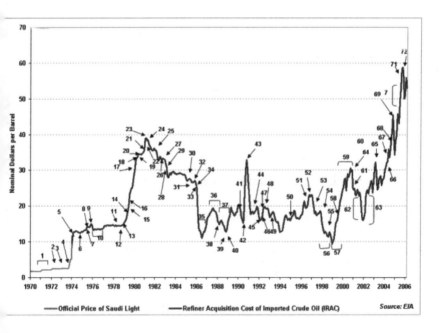

The price data graphed above are in nominal terms, i.e., they are in dollars-of-the-day and have not been adjusted for inflation. Historical and forecast real and nominal crude oil and gasoline price information is maintained on a more frequent basis on the Short Term Energy Outlook Webpage.

1. OPEC begins to assert power; raises tax rate & posted prices

2. OPEC begins nationalization process; raises prices in response to falling US dollar

3. Negotiations for gradual transfer of ownership of western assets in OPEC countries

4. Oil embargo begins (October 19-20, 1973)

5. OPEC freezes posted prices; US begins mandatory oil allocation

6. Oil embargo ends (March 18, 1974)

7. Saudis increase tax rates and royalties

8. US crude oil entitlements program begins

9. OPEC announces 15% revenue increase effective October 1, 1975

10. Official Saudi Light price held constant for 1976

Appendix II

11. Iranian oil production hits a 27-year low

12. OPEC decides on 14.5% price increase for 1979

13. Iranian revolution; Shah deposed

14. OPEC raises prices 14.5% on April 1, 1979

15. US phased price decontrol begins

16. OPEC raises prices 15%

17. Iran takes hostages; President Carter halts imports from Iran; Iran cancels US contracts; Non-OPEC output hits 17.0 million b/d

18. Saudis raise marker crude price from $19/bbl to $26/bbl

19. Windfall Profits Tax enacted in U.S.

20. Kuwait, Iran, and Libya production cuts drop OPEC oil production to 27 million b/d

21. Saudi Light raised to $28/bbl

22. Saudi Light raised to $34/bbl

23. First major fighting in Iran-Iraq War

24. President Reagan abolishes remaining price and allocation controls

25. Spot prices dominate official OPEC prices

26. US boycotts Libyan crude; OPEC plans 18 million b/d output

27. Syria cuts off Iraqi pipeline

28. Libya initiates discounts; non-OPEC output reaches 20 million b/d; OPEC output drops to 15 million b/d

29. OPEC cuts prices by $5/bbl and agrees to 17.5 million b/d output – January 1983

30. Norway, United Kingdom, and Nigeria cut prices

31. OPEC accord cuts Saudi Light price to $28/bbl

32. OPEC output falls to 13.7 million b/d

33. Saudis link to spot price and begin to raise output in June 1985

34. OPEC output reaches 18 million b/d

35. Wide use of netback pricing

36. Wide use of fixed prices

37. Wide use of formula pricing

38. OPEC/Non-OPEC meeting failure

Appendix II

39. OPEC production accord; Fulmar/Brent production outages in the North Sea

40. Exxon's Valdez tanker spills 11 million gallons of crude oil

41. OPEC raises production ceiling to 19.5 million b/d – June 1989

42. Iraq invades Kuwait

43. Operation Desert Storm begins; 17.3 million barrels of SPR crude oil sales is awarded

44. Persian Gulf war ends

45. Dissolution of Soviet Union; last Kuwaiti oil fire is extinguished on November 6, 1991

46. UN threaten sanctions against Libya

47. Saudi Arabia agrees to support OPEC price increase

48. OPEC production reaches 25.3 million b/d, the highest in over a decade

49. Kuwait boosts production by 560,000 b/d in defiance of OPEC quota

50. Nigerian oil workers' strike

51. Extremely cold weather in the US and Europe

52. U.S. launches cruise missile attacks into southern Iraq following an Iraqi-supported invasion of Kurdish safe haven areas in northern Iraq

53. Iraq begins exporting oil under United Nations Security Council Resolution 986

54. Prices rise as Iraq's refusal to allow United Nations weapons inspectors into sensitive sites raises tensions in the oil-rich Middle East

55. OPEC raises its production ceiling by 2.5 million barrels per day to 27.5 million barrels per day, this is the first increase in 4 years

56. World oil supply increases by 2.25 million barrels per day in 1997, the largest annual increase since 1988

57. Oil prices continue to plummet as increased production from Iraq coincides with no growth in Asian oil demand due to the Asian economic crisis and increases in world oil inventories following two unusually warm winters

58. OPEC pledges additional production cuts for the third time since March 1998; total pledged cuts amount to about 4.3 million barrels per day

59. Oil prices triple between January 1999 and September 2000 due to strong world oil demand,

OPEC oil production cutbacks, and other factors, including weather and low oil stock levels

60. President Clinton authorizes the release of 30 million barrels of oil from the Strategic Petroleum Reserve (SPR) over 30 days to bolster oil supplies, particularly heating oil in the Northeast

61. Oil prices fall due to weak world demand (largely as a result of economic recession in the United States) and OPEC overproduction

62. Oil prices decline sharply following the September 11, 2001 terrorist attacks on the United States, largely on increased fears of a sharper worldwide economic downturn (and therefore sharply lower oil demand); prices then increase on oil production cuts by OPEC and non-OPEC at the beginning of 2002, plus unrest in the Middle East and the possibility of renewed conflict with Iraq

63. OPEC oil production cuts, unrest in Venezuela, and rising tension in the Middle East contribute to a significant increase in oil prices between January and June

64. A general strike in Venezuela, concern over a possible military conflict in Iraq, and cold winter weather all contribute to a sharp decline in U.S. oil inventories and cause oil prices to escalate further at the end of the year

65. Continued unrest in Venezuela and oil traders' anticipation of imminent military action in Iraq causes prices to rise in January and February 2003

66. Military action commences in Iraq on March 19, 2003; Iraqi oil fields are not destroyed as had been feared; prices fall

67. OPEC delegates agree to lower the cartel's output ceiling by 1 million barrels per day, to 23.5 million barrels per day, effective April 2004

68. OPEC agrees to raise its crude oil production target by 500,000 barrels (2% of current OPEC production) by August 1 in an effort to moderate high crude oil prices

69. Hurricane Ivan causes lasting damage to the energy infrastructure in the Gulf of Mexico and interrupts oil and natural gas supplies to the United States; U.S. Secretary of Energy Spencer Abraham agrees to release 1.7 million barrels of oil in the form of a loan from the SPR

70. Continuing oil supply disruptions in Iraq and Nigeria, as well as strong energy demand, raise prices during the first and second quarters of 2005

71. Tropical Storm Cindy and Hurricanes Dennis,

Appendix II

Katrina, and Rita disrupt oil supply in the Gulf of Mexico

72. President Bush authorizes SPR release

Original concept for the chart was by the Analysis Division in the Office of Management Operations, SPR Modified and updated by the Office of Energy Markets and End Use in the Energy Information Administration.

Index

A
Afghanistan, 49
Alaska Pipeline, 53
Alberta, 81-82
al-Gaddafi, Muammar, 40
Algeria, 102, 149
Al-Naimi, Ali I., 93
Anglo-Persian Oil Company, 35
Arabian Light, 24
Arab-Israeli War, 25
Arab Oil Embargo, 25, 135

B
Big Oil, 28, 38
biomass, 77, 135, 137
British Petroleum, 30, 35, 37
Bretton Woods system, 43
Bush, President George W., 26, 113, 127

C
Cairn Energy, 64

California, 9, 30, 59, 60, 85, 105-107, 133, 148
Cardenas, Lazaro, 33
Carter, President Jimmy, 50, 91, 121
China, 61-62, 71-72, 74, 78, 89, 97, 117
CNN Presents, 73
coal bed methane (CBM), 98
Cushing, Oklahoma, 83

D
Department of Energy (DOE), 13, 15, 19, 55-56, 59, 83, 85, 119
Directorate General of Hydrocarbons (DGH), 67

E
Ecuador, 149
Eisenhower, President Dwight D., 36-37
Energy Advocates, 109-110,
Energy Information Administration (EIA), 12, 119, 127

F
Fifty/Fifty Principle, 35
FutureGen, 85

G
Gabon, 149
GDP Per Capita, 61-62
Ghawar, 22-24
Goldman Sachs, 71

Gulf of Mexico, 73, 126-127
Gulf Oil, 30
Gulf War, 28, 123

H
Highlands Ranch, Colorado, 57
Hubbert, Dr. M. King, 19
Hydrocarbon Age, 87, 90
Hydrocarbon Vision 2025, 66-67, 70

I
India, 15, 61-70, 72, 74, 89, 97, 117
India Brand Equity Foundation (IBEF), 66
Indonesia, 149
International Energy Agency (IEA), 19, 61
International Society of Energy Advocates, 75, 109
Iran, 34-35, 44, 49-50, 68, 98-103, 121

J
Japan, 41

K
Kazakestan, 102
Khomeini, Ayatollah, 49
The Kingdom, 73
Kingdom of Saudi Arabia, 22, 24, 88

L
Libya, 40, 68, 121-123, 149

Liquified Natural Gas (LNG), 95

M
"Malaise Speech", 50
Managa Field, 64
Mattei, Enrico, 31
Maxwell, Charley, 6, 90
Mexico, 33-34, 61,
Middle East, 21, 25-26, 28, 42-44, 49, 71, 73, 88-90, 102, 108, 124-125,
Mossadegh, Mohammed, 35

N
natural gas, 8-9, 13-14, 136, 138-139, 142, 144, 147-148, 150-153
New Exploration Licensing Program (NELP), 64, 68
Nigeria, 11, 122, 126, 149
Nixon, President Richard M., 43, 45-48
North Sea, 48, 89, 123

O
OAPEC, 44, 46-48
oil consumption, 59, 105
Oil Sands, 81-83
Oil Storm, 73
ONGC, 64, 68
OPEC, 8, 21, 28-41, 43-44, 89-93, 102-103, 109, 120-126

P

Pahlavi, HRH Mohammad Reza, 49
Pahlavi, HRH Reza, 6, 99
Peak Oil, 17, 19
Petro-State, 41
petroleum security, 8, 28, 48, 52, 73, 76, 105
The Prize, 52
Proposition 87, 9, 105
Prudhoe Bay, 53

Q

Qatar, 102

R

renewable/alternative energies, 77, 83
Royal Dutch Petroleum/Shell Oil, 30
Russia, 9, 68, 101, 102

S

Sakhalin, 68
Saudi Aramco, 22
Seven Sisters, 31-36, 38-39
Sierra Club, 53
Silicon Valley, 59
Simmons, Matt, 21-22
Six Days War, 44
Socony-Vacuum (now part of Exxon-Mobil), 30

Soviet Union, 49, 123
Standard Oil Company, 30
Standard Oil of California (now Chevron), 30
Standard Oil of New Jersey (now part of Exxon-Mobil), 30
Strategic Petroleum Reserve, 56, 119, 125
Syriana, 73

T
Texaco, 30
Twilight in the Desert, 21

U
United Arab Emirates, 149
University of Oklahoma, 110

V
Venezuela, 35-36, 39, 68, 82, 125-126, 149

Y
Yergin, Daniel, 52
Yom Kippur War, 41, 44

Glossary

A

3-D Seismic - A relatively new exploration technique used in the search for oil and gas underground structures. The basic premise behind seismic is the same as ultrasound technology used in the medical field. Sound from a shot hole is recorded from geophones and interpreted to give a picture of the underlying structures within the earth. 3-D has now become a common practice to redefine and identify known as well as unknown structures. Many times these structures contain traps that hold oil and gas yet to be discovered.

4-D Seismic - The newest advances in seismic technology which now takes into consideration a fourth dimension, time. With 4-D seismic, geologists are now able to monitor the movement and the mobility of oil as it is extracted in the production process.

Glossary

Anticline - A geological term describing a fold in the earth's surface with strata sloping downward on both sides from a common crest. Anticlines frequently have surface manifestations like hills, knobs, and ridges. At least 80 percent of the world's oil and gas has been found in anticlines.

Arab Oil Embargo of 1973-74 - During the Arab-Israeli conflict in October 1973, Arab oil producers cut off shipments to the United States and the Netherlands in retaliation for their support of Israel. At the same time, they cut down production. The shortage was felt by all oil-importing nations, with world prices moving sharply higher. Price and allocation controls suppressed some of this increase in the United States, but gasoline lines were still prevalent.

B

Basin - A depression in the earth's crust in which sedimentary materials have accumulated. Such a basin may contain oil or gas fields.

BCF (billion cubic feet) - The cubic foot is a standard unit of measure for gas at atmospheric pressure.

Biomass - Any organic material, such as wood, plants, and organic wastes, that can be turned into fuel.

Btu (British thermal unit) - A standard measure of heat content in a fuel. One Btu equals the amount of energy required to raise the temperature of one pound of water one degree Fahrenheit at or near 39.2 degrees Fahrenheit.

Butane - A hydrocarbon associated with petroleum. It is gaseous at ordinary atmospheric conditions.

C

Christmas tree - An assembly of valves, gauges, and chokes mounted on a well casinghead to control production and the flow of oil to the pipelines.

Clean oil - Crude oil containing less than 1 percent sediment and water; also called pipeline oil—oil clean enough to send through a pipeline.

CO2 injection - A secondary recovery technique in which carbon dioxide (CO_2) is injected into wells as part of a miscible recovery program.

Coal gasification - The chemical conversion of coal to synthetic gaseous fuel.

Coal liquefaction - The chemical conversion of coal to synthetic liquid fuel.

Completed well - A well made ready to produce oil or natural gas. Completion involves cleaning out the well, running steel casing and tubing into the hole, adding permanent surface control equipment, and perforating the casing so oil or gas can flow into the well and be brought to the surface.

Conventional energy sources - Oil, gas, coal, and sometimes nuclear energy, in contrast to alternative energy sources such as solar, hydroelectric and geothermal power, synfuels, and biomass.

Cracking - The process of breaking down the larger, heavier and more complex hydrocarbon molecules into simpler and lighter molecules, thus increasing the gasoline yield from crude oil. Cracking is done by application of heat and pressure, and in modern time the use of a catalytic agent.

Crude oil - Liquid petroleum as it comes out of the ground. Crude oils range from very light (high in gasoline) to very heavy (high in residual oils). Sour crude is high in sulfur content. Sweet crude is low in sulfur and therefore often more valuable.

Crude oil equivalent - A measure of energy content that converts units of different kinds of energy into the energy equivalent of barrels of oil.

D

Demand destruction - The reduction of demand for a commodity as a result of high prices.

Diesel oil - A petroleum fraction composed primarily of aliphatic (linear of unbranched) hydrocarbons. Diesel oil is slightly heavier than kerosene.

Directional drilling - Drilling at an angle, instead of on the perpendicular, by using a whipstock to bend the pipe until it is going in the desired direction. Directional drilling is used to develop offshore leases, where it is very costly and sometimes impossible to prepare separate sites for every well; to reach oil beneath a building or some other location which cannot be drilled directly; or to control damage or as a last resort when a well has cratered. It is much more expensive than conventional drilling procedures.

Distillate - Liquid hydrocarbons, usually colorless and of high API gravity, recovered from wet gas by a separator that condenses the liquid out of the gas. The present term is natural gas.

Distillate fuel oil - A term subject to a variety of definitions. Sometimes the definition is based on the method of production, but other definitions are based on boiling range, viscosity, or use.

Distributor - A wholesaler of gasoline and other petroleum products; also known as a jobber. Distributors of natural gas are almost always regulated utility companies.

Domestic production - Oil and gas produced in the United States as opposed to imported product.

Downstream - All operations taking place after crude oil is produced, such as transportation, refining, and marketing.

Drilling - The act of boring a hole through which oil or gas may be produced if encountered in commercial quantities.

Dry hole - A well that either produces no oil or gas or yields too little to make it economic to produce.

Dry natural gas - Natural gas containing few or no natural gas liquids (liquid petroleum mixed with gas).

Dual completion - Completing a well that draws from two or more separate producing formations at different depths. This is done by inserting multiple strings of tubing into the well casing and inserting packers to seal off all formations except the one to be produced by a particular string.

DWT - dead weight tons

E

Enhanced oil recovery - Injection of water, steam, gases or chemicals into underground reservoirs to cause oil to flow toward producing wells, permitting more recovery than would have been possible from natural pressure or pumping alone.

Ethanol - The two-carbon-atom alcohol present in the greatest proportion upon fermentation of grain and other renewable resources such as potatoes, sugar, or timber. Also called grain alcohol.

Exploration - The search for oil and gas. Exploration operations include: aerial surveys, geophysical surveys, geological studies, core testing and the drilling of test wells.

Exploratory well - A well drilled to an unexplored depth or in unproven territory, either in search of a new reservoir or to extend the known limits of a field that is already partly developed.

Extraction plant - A plant for the extraction of the liquid constituents in casinghead gas or wet gas.

F

Fault - A break in the continuity of stratified rocks or even basement rocks. Faults are significant to oilmen because they can form traps for oil when the rock fractures, they can break oil reservoirs into noncommunicating sections, they help produce oil accumulations, and they form traps on their own.

Fault trap - A geological formation in which oil or gas in a porous section of rock is sealed off by a displaced, nonporous layer.

Field - A geographical area under which one or more oil or gas reservoirs lie, all of them related to the same geological structure.

Five-spot waterflood program - A secondary-recovery operation in which four injection wells are drilled in a square pattern with the production well in the center. Water from the injection wells moves through the formation, forcing oil toward the production well.

Flooding - One of the methods of enhanced oil recovery. Water flooding or gas flooding might be considered secondary recovery methods.

Flowing well - A well that produces through natural reservoir pressure and does not require pumping.

Formation - A geological term that describes a succession of strata similar enough to form a distinctive geological unit useful for mapping or description.

Fossil fuels - Fuels that originate from the remains of living things, such as coal, oil, natural gas, and peat.

Fracturing - A well stimulation technique in which fluids are pumped into a formation under extremely high pressure to create or enlarge fractures for oil and gas to flow through. Proppants such as sand are injected with the liquid to hold the fractures open.

Fuel oil - See Heating oil.

G

Gas cap - The gas that exists in a free state above the oil in a reservoir.

Gas condensate - Liquid hydrocarbons present in casinghead gas that condense when brought to the surface.

Gas lift - A recovery method that brings oil from the bottom of a well to the surface by using compressed gas. Gas pumped to the bottom of the reservoir mixes with fluid, expands it, and lifts it to the surface.

Gas-oil ratio - The number of cubic feet of natural gas produced along with a barrel of oil.

Gasoline - A volatile, inflammable, liquid hydrocarbon mixture.

Geophones - The sound-detecting instruments used to measure sound waves created by explosions set off during seismic exploration work.

Geophysicist - A geophysicist applies the principles of physics to the understanding of geology.

Geothermal energy - Energy produced from subterranean heat.

Gun perforation - A method of creating holes in a well casing downhole by exploding charges to propel steel projectiles through the casing wall. Such holes allow oil from the formation to enter the well.

Gusher - A well drilled into a formation in which the crude is under such high pressure that at first it spurts out of the wellhead like a geyser. Gushers are rare today owing to improved drilling technology, the use of drilling mud to control downhole pressure, and oilmen's recognition of their wastefulness.

H

Heating oil - Oil used for residential heating.

Heavy oil - A type of crude petroleum characterized by high viscosity and a high carbon-to-hydrogen ratio. It is usually difficult and costly to produce by conventional techniques.

Horizontal drilling - The newer and developing technology that makes it possible to drill a well from the surface, vertically down to a certain level, and then to turn at a right angle, and continue drilling horizontally within a specified reservoir, or an interval of a reservoir.

Hydraulic fracturing - A method of stimulating production from a low-permeability formation by creating fractures and fissures by applying very high fluid pressure.

Hydrocarbons - A large class of organic compound of hydrogen and carbon. Crude oil, natural gas, and natural gas condensate are all mixtures of various hydrocarbons, among which methane is the simplest.

I

In situ - In its original place. Refers to methods of producing synfuels underground, such as underground gasification of a coal seam or heating oil shale underground to release its oil.

Independent producer - 1. A person or corporation

that produces oil for the market, who has no pipeline system or refining capability. 2. An oil entrepreneur who secures financial backing and drills his own wells.

Injection well - A well employed for the introduction of water, gas, or other fluid under pressure into an underground stratum. Injection wells are employed for the disposal of salt water produced with oil or other waste. They are also used for a variety of other purposes: 1) Pressure maintenance, to introduce a fluid into a producing formation to maintain underground pressures which would otherwise be reduced by virtue of the production of oil or gas, 2) Secondary recovery operations, to introduce a fluid to decrease the viscosity of oil, reduce its surface tension, lighten its specific gravity, and drive oil into producing wells, resulting in greater production of oil.

J

Jack-up rig - A floating platform with legs on each corner that can be lowered to the sea bottom to raise or jack the platform above the water.

Joint Operating Agreement - A detailed written agreement between the working interest owners of a property which specifies the terms according to which that property will be developed.

Joint venture - A large-scale project in which two or more parties (usually oil companies) cooperate. One supplies funds and the other actually carries out the project. Each participant retains control over his share, including liability and the right to sell.

K

Kerogen - The hydrocarbon in oil shale. Scientists believe that kerogen was the precursor of petroleum and that petroleum development in shale was somehow prematurely arrested.

Kerosene - The petroleum fraction containing hydrocarbons that are slightly heavier than those found in gasoline and naphtha. Kerosene was the most important petroleum product because of its use for home and commercial lighting; in recent years demand has risen again as a result of kerosene's use in gas turbines and jet engines.

L

Limestone - Sedimentary rock largely consisting of calcite. On a world-wide scale, limestone reservoirs probably contain more oil and gas reserves than all other types of reservoir rock combined.

LNG (liquefied natural gas) - Natural gas that has been converted to a liquid through cooling to -260 degrees Fahrenheit at atmospheric pressure.

LPG (liquefied petroleum gases) - Hydrocarbon fractions lighter than gasoline, such as ethane, propane, and butane, kept in a liquid state through compression and/or refrigeration, commonly referred to as bottled gas.

M

Mid-continent crude - Oil produced mainly in Kansas, Oklahoma, and North Texas.

Midstream or middle distillates - Refinery products in the middle of the distillation range of crude oil, including kerosene, kerosene-based jet fuel, home heating fuel, range oil, stove oil and diesel fuel.

MCF (Million cubic feet.) - The cubic foot is a standard unit of measure for quantities of gas at atmospheric pressure.

Mud - A fluid mixture of clay, chemicals, and weighting materials suspended in fresh water, salt water, or diesel oil.

N

Natural gas - A mixture of hydrocarbon compounds

and small amounts of various nonhydrocarbons (such as carbon dioxide, helium, hydrogen sulfide, and nitrogen) existing in the gaseous phase or in solution with crude oil in natural underground reservoirs.

Naval petroleum reserves - Areas containing proven oil reserves that were set aside for national defense purposes by Congress in 1923 (located in Elk Hills and Buena Vista, California; Teapot Dome, Wyoming; and on the North Slope in Alaska).

NGL (natural gas liquids) - Portions of natural gas that are liquefied at the surface in lease separators, field facilities, or gas processing plants, leaving dry natural gas. They include, but are not limited to, ethane, propane, butane, natural gasoline, and condensate.

O

Offshore platform - A fixed structure from which wells are drilled offshore for the production of oil and natural gas.

Oil rig - A drilling rig that drills for oil and gas.

Oil shale - A fine-grained, sedimentary rock that contains kerogen, a partially formed oil. Kerogen can be extracted by heating the shale, but at a very high cost.

OPEC (Organization of Petroleum Exporting Countries) - An international oil cartel originally formed in 1960 and including in 1983: Saudi Arabia, Kuwait, Iran, Iraq, Venezuela, Quatar, Libya, Indonesia, United Arab Emirates, Algeria, Nigeria, Ecuador, and Gabon.

Operator - The individual or company responsible for the drilling, completion and production operations of a well, and the physical maintenance of the leased property.

P

Perforating gun - An instrument lowered at the end of a wireline into a cased well. It contains explosive charges that can be electronically detonated from the surface.

Perforation - A method of making holes through the casing opposite the producing formation to allow the oil or gas to flow into the well. See *Gun perforation.*

Permeability - A measure of the ease with which a fluid such as water or oil moves through a rock when the pores are connected. Geologists express permeability in a unit named the darcy, but oilmen use the millidarcy because most of the rocks they come in contact with are not very permeable.

Petrochemicals - Chemicals derived from crude oil or natural gas, including ammonia, carbon black, and other organic chemicals.

Petroleum - Strictly speaking, crude oil. Also used to refer to all hydrocarbons, including oil, natural gas, natural gas liquids, and related products.

Petroleum engineer - A term including three areas of specialization: 1) drilling engineers specialize in the drilling, workover, and completion operations, 2) production engineers specialize in studying a well's characteristics and using various chemical and mechanical procedures to maximize the recovery from the well, 3) reservoir engineers design and execute the planned development of a reservoir. Many U.S. universities offer BS, MS, and PhD degrees in petroleum engineering.

Petroleum geologist - A geologist who specializes in the exploration for, and production of, petroleum.

Pipeline - A tube or system of tubes used for the transportation of oil or gas. Types of oil pipelines include: lead lines, from pumping well to a storage tank; flow lines, from flowing well to a storage tank; lease lines, extending from the wells to lease tanks; gathering lines, extending from lease tanks to a central accumulation point; feeder lines, extending from leases to trunk lines; and trunk lines, extending from a producing area to refineries or terminals.

Pipeline gas - Gas under enough pressure to enter the high-pressure gas lines of a purchaser; gas in which enough liquid hydrocarbons have been removed so that such liquids will not condense in the transmission lines.

Plugging a well - Filling the borehole of an abandoned well with mud and cement to prevent the flow of water or oil from one strata to another or to the surface.

Porosity - A measure of the number and size of the spaces between each particle in a rock. Porosity affects the amount of liquid and gases, such as natural gas and crude oil, that a given reservoir can contain.

Possible reserves - Areas in which production of crude oil is presumed possible owing to geological inference of a strongly speculative nature.

Primary recovery - Production in which oil moves from the reservoir, into the wellbore, under naturally occurring reservoir pressure.

Production - A term commonly used to describe

taking natural resources out of the ground.

Prospect - A lease or group of leases on which an operator intends to drill.

Proved behind-pipe reserves - Estimates of the amount of crude oil or natural gas recoverable by recompleting existing wells.

Proved developed reserves - Estimates of what is recoverable from existing wells with existing facilities from open, producing payzones.

Proved reserves - Estimates of the amount of oil or natural gas believed to be recoverable from known reservoirs under existing economic and operating conditions.

Proved undeveloped reserves - Estimates of what is recoverable through new wells on undrilled acreage, deepening existing wells, or secondary recovery methods.

Pump - A device that is installed inside or on a production string (tubing) that lifts liquids to the surface.

Pumping well - A well that does not flow naturally and requires a pump to bring product to the surface.

Q
Quad - One Quadrillion Btus

R
Recoverable resources - An estimate of resources, including oil and/or natural gas, both proved and undiscovered, that would be economically extractable under specified price-cost relationships and technological conditions.

Refining - Manufacturing petroleum products by a series of processes that separate crude oil into its major components and blend or convert these components into a wide range of finished products, such as gasoline or jet fuel.

Reserve - That portion of the identified resource from which a usable mineral and energy commodity can be economically and legally extracted at the time of determination.

Reservoir - A porous, permeable sedimentary rock formation containing quantities of oil and/or gas enclosed or surrounded by layers of less permeable or impervious rock. Also called a *horizon*.

Reservoir pressure - The pressure at the face of the producing formation when the well is shut in. It equals the shut-in pressure at the wellhead plus the weight of the column of oil in the hole.

Rotary drilling - A method of well-drilling that employs a rotating bit and drilling mud to cut through rock formations.

S

Secondary recovery - The introduction of water or gas into a well to supplement the natural reservoir drive and force additional oil to the producing wells.

Sedimentary basin - A large land area composed of unmetamorphized sediments. Oil and gas commonly occur in such formations.

Sedimentary rock - Rock formed by the deposition of sediment, usually in a marine environment.

Seismic exploration - A method of prospecting for oil or gas by sending shock waves into the earth. Different rocks transmit, reflect, or refract sound waves at different speeds, so when vibrations at the surface send sound waves into the earth in all directions, they reflect to the surface at a distance and angle from the sound source that indicates the depth of the interface. These reflections are recorded and analyzed to map underground formations.

Seismograph - A device that records natural or

manmade vibrations from the earth. Geologists read what it has recorded to evaluate the oil potential of underground formations.

Shale A - type of rock composed of common clay or mud.

Shale oil - The substance produced from the treatment of kerogen, that hydrocarbon found in some shales, which is difficult and costly to extract. About 34 gallons of shale oil can be extracted from one ton of ore.

Stratigraphic trap - A porous section of rock surrounded by nonporous layers, holding oil or gas. They are usually very difficult to locate, although oilmen believe that most of the oil yet to be discovered will be found in these traps.

Structural trap - A reservoir created by some cataclysmic geologic event that creates a barrier and prevents further migration. The most common structural traps are anticlines, in which at least 80 percent of the world's oil and gas have been discovered.

Structure - Subsurface folds or fractures of strata that form a reservoir capable of holding oil or gas.

Sweet crude - Crude oil with low sulfur content which is less corrosive, burns cleaner, and requires less processing to yield valuable products.

Syncline - A downfold in stratified rock that looks like an upright bowl. Unfavorable to the accumulation of oil and gas.

Synfuels - Fuels produced through chemical conversions of natural hydrocarbon substances such as coal and oil shale.

Synthetic crude oil (syncrude) - A crude oil derived from processing carbonaceous material such as shale oil or unrefined oil in coal conversion processes.

Synthetic gas - Gas produced from solid hydrocarbons such as coal, oil shale, or tar sands.

T

Take-or-pay contract - A long-term contract between a gas producer and a gas purchaser, such as a pipeline transmission company.

Tank bottoms - A mixture of oil, water, and other foreign matter that collects in the bottoms of stock tanks and large crude storage tanks and must be cleaned or pumped out on a regular basis.

Tanker - An ocean-going ship which hauls crude oil.

Tar sands - Rocks other than coal or oil shale that contain highly viscous hydrocarbons that are unrecoverable by primary production methods.

TCF - Trillion cubic feet.

Tertiary recovery - The recovery of oil that involves complex and very expensive methods such as the injection of steam, chemicals, gases, or heat, as compared to primary recovery, which involves depleting a naturally flowing reservoir, or secondary recovery, which usually involves repressuring or waterflooding.

Therm - A measure of heat content. One therm equals 100,000 Btus.

Tight formation - A sedimentary layer of rock cemented together in a manner that greatly hinders the flow of any gas through the rock.

Tight hole - A well about which the operator keeps all information secret.

Tight Sand - A formation with low permeability. Gas produced from a formation so designated by the Federal Energy Regulatory Commission qualifies for a higher market price.

Total depth (TD) - The maximum depth of a borehole.

Trap - A natural configuration of layers of rock where non-porous or impermeable rocks act as a barrier, blocking the natural upward flow of hydrocarbons.

U

ULCC (Ultralarge crude carrier) - A large tanker built especially to carry 500,000 dwt and up of crude oil.

Undiscovered recoverable resources - Resources outside of known fields, estimated from broad geologic knowledge and theory.

Upstream - Activities concerned with finding petroleum and producing it, compared to downstream which are all the operations that take place after production.

V

Viscosity - A fluid's resistance to flowing.

VLCC (very large crude carrier) - A tanker built to carry 200,000 to 350,000 dwt of crude oil.

W

Waterflooding - A secondary recovery method in which water is injected into a reservoir to force additional oil into the wells.

Wellbore - Physically, wellbore refers to a borehole, in other words a completed well.

Wellhead - A device on the surface used to hold the tubing in the well. The wellhead is the originating point of the producing well at the top of the ground.

Workover - To clean out or work on a well to restore or increase production.

Workover rig - The rig used when oilmen try to restore or increase a well's production.

X, Y, Z

Zone - A specific interval of rock strata containing one or more reservoirs, used interchangeably with *formation*.

Disclaimer: This glossary is believed to have been originated by Larry C. Neely of Maverick Energy Inc. Various version of this glossary have circulated around the Internet, and Mr. Neely is often uncredited. We have attempted to contact Mr. Neely with no success, but he has stated on his website that this glossary can be republished. Portions of this book were taken from notes generated by Jason Reimbold when researching energy markets for Ensaga Energy, LLC. The discussion of energy issues and policy rely on data drawn almost entirely from publicly available sources. It is our intention to name all sources used for the writing of this book.

Energy Conversion Factors

ENERGY TYPE	BTU/BARREL	BTU/GALLON
CRUDE OIL	5,855,795	139,424
MOTOR GASOLINE	5,250,000	125,000
AVIATION GASOLINE	5,005,224	119,172
JET FUEL	5,434,926	129,403
L.P.G.	4,054,470	96,535
PROPANE	3,836,000	91,333
ETHANE	3,082,000	73,381
BUTANE	4,326,000	103,000
KEROSENE	5,670,000	135,000
#1 DISTILLATE	5,706,000	135,857
#2 DISTILLATE	5,825,000	138,690
#4 DISTILLATE	6,062,000	144,333
RESIDUAL OIL	6,287,000	149,690
	BTU/TON	**BTU/POUND**
ANTHRACITE COAL	25,400,000	12,700
BITUMINOUS COAL	23,750,000	11,875
	BTU/KWH	
ELECTRICITY (END-USERS)	3,412	
ELECTRICITY (GENERATION)	10,908	
	BTU/CUBIC FOOT	
NATURAL GAS	1,031	
	BTU/CORD	
WOOD (OAK)	22,750,000	
WOOD (MAPLE)	21,850,000	
		BTU/POUND
REFUSE DERIVED FUEL (GOOD)		7,200
REFUSE DERIVED FUEL (AVERAGE MASS BURN)		5,500

- 1 BARREL = 42 GALLONS
- 1 BARREL OF CRUDE OIL = 0.150 SHORT TONS = 0.136 METRIC TONS
- 1 SHORT TON = 2,000 POUNDS = 6.65 BARRELS OF CRUDE OIL
- 1 METRIC TON = 2,204.62 POUNDS = 7.33 BARRELS OF CRUDE OIL
- 1 METRIC TON OF GASOLINE = 345 GALLONS
- 1 METRIC TON OF HEATING OIL = 312 GALLONS
- 1 METRIC TON OF RESIDUAL OIL = 6.5 BARRELS
- 1 CCF = 100 CUBIC FEET
- 1 MCF = 1,000 CUBIC FEET
- 1 THERM = 100,000 BTU = APPROX. 100 CUBIC FEET OF NATURAL GAS

Source: State of Connecticut

About the Authors

Mark A. Stansberry

Mark A. Stansberry is Chairman of The GTD Group (www.thegtdgroup.com) and has been active in the energy industry for the past thirty years. He has been active in banking, real estate development, international government relations and business ventures.

He is President of The International Society of The Energy Advocates, Chairman of Oklahoma's State Chamber Energy Council, and Chairman and Founder of the International Energy Policy Conference.

His is a graduate of Oklahoma Christian University, and his memberships include Oklahoma Independent Petroleum Association, Independent Petroleum Association of America, Texas Alliance of Energy Producers, and the Association of International Petroleum Negotiators. He has testified before the U. S. Senate Energy and Natural Resources Committee along with other regulatory bodies, and he served on the staff of former U. S. Senator Dewey F. Bartlett, 1975-76. In 1991, he co-authored the handbook entitled, *The Acquisition Process and Due*

Diligence: Minimize Risk & Maximize Return!

He was appointed by Oklahoma Governor Frank Keating to serve as a member of the Board of Regents of the Regional University System of Oklahoma in 2001 and has served as Chairman. In addition, he has served as Chairman of the Governor's International Team, Board of Directors of People to People International, Board of Trustees of The Fund for American Studies, Board of Trustees for Oklahoma Christian University, Board of Directors of Oklahoma Heritage Association, and serves on other boards and advisory committees.

Former U. S. Senator and Governor Henry Bellmon of Oklahoma stated in 2001 that Mark is "Oklahoma's Energy Authority."

Jason P. Reimbold

Jason Reimbold is a Vice President for *the Theseus Group* (www.thetheseusgroup.com), an energy M&A consulting firm. Before joining this company, Jason was an analyst for the Bank of Oklahoma Energy Group. He graduated with a degree in finance from the University of Tulsa where he studied geopolitics and international energy markets under former Deputy Assistant Secretary of Energy, R. Dobie Langenkamp.

From 1999 to 2001, Jason served an extended tour in the Republic of South Korea while in the U.S. Army Cavalry. After returning to the United States in early 2001, he was staffed as a Battalion Intelligence Officer at Fort Drum, New York. In the midst of 9/11, Jason was responsible for conducting battalion level anti-terrorism training and security briefings. This duty began his research of the Middle-East and global energy politics.

In 2005, Jason founded www.GlobalOilWatch.com which has become a leading energy research portal for industry analysts and investors. He also serves as the Vice President of Membership (Ex-Officio) for Energy Advocates.